how to
know the
amphibians
and reptiles

The **Pictured Key Nature Series** has been published since 1944 by the Wm. C. Brown Company. The series was initiated in 1937 by the late Dr. H. E. Jaques, Professor Emeritus of Biology at Iowa Wesleyan University. Dr. Jaques' dedication to the interest of nature lovers in every walk of life has resulted in the prominent place this series fills for all who wonder "**How to Know.**"

John F. Bamrick and Edward T. Cawley
Consulting Editors

The Pictured Key Nature Series

How to Know the
  AMPHIBIANS AND REPTILES, Ballinger-Lynch
  AQUATIC INSECTS, Lehmkuhl
  AQUATIC PLANTS, Prescott, Second Edition
  BEETLES, Arnett-Downie-Jaques, Second Edition
  BUTTERFLIES, Ehrlich
  FALL FLOWERS, Cuthbert
  FERNS AND FERN ALLIES, Mickel
  FRESHWATER ALGAE, Prescott, Third Edition
  FRESHWATER FISHES, Eddy-Underhill, Third
    Edition
  FRESHWATER CRUSTACEA, Fitzpatrick
  GILLED MUSHROOMS, Smith-Smith-Weber
  GRASSES, Pohl, Third Edition
  IMMATURE INSECTS, Chu
  INSECTS, Bland-Jaques, Third Edition
  LICHENS, Hale, Second Edition
  LIVING THINGS, Winchester-Jaques, Second
    Edition
  MAMMALS, Booth, Fourth Edition
  MITES AND TICKS, McDaniel
  MOSSES AND LIVERWORTS, Conard-Redfearn,
    Second Edition
  NON-GILLED MUSHROOMS, Smith-Smith-Weber,
    Second Edition
  PLANT FAMILIES, Jaques, Second Edition
  POLLEN AND SPORES, Kapp
  PROTOZOA, Jahn, Bovee, Jahn, Second Edition
  SEAWEEDS, Abbott-Dawson, Second Edition
  SEED PLANTS, Cronquist
  SPIDERS, Kaston, Third Edition
  SPRING FLOWERS, Verhoek-Cuthbert, Second
    Edition
  TREES, Miller-Jaques, Third Edition
  TRUE BUGS, Slater-Baranowski
  TRUE SLIME MOLDS, Farr
  WEEDS, Wilkinson-Jaques, Third Edition
  WESTERN TREES, Baerg, Second Edition

how to
know the
amphibians
and reptiles

Royce E. Ballinger
University of Nebraska

John D. Lynch
University of Nebraska

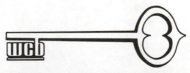

The Pictured Key Nature Series
Wm. C. Brown Company Publishers
Dubuque, Iowa

2 04786 01

Contents

Preface

Until now the identification of amphibians and reptiles has depended on various regional or specialized manuals or field guides of which there are many of varying quality and availability. Alternatively one could utilize Blair et al. (1968) *Vertebrates of the United States* which contained the only identification keys for the entire U.S. Nomenclatural changes and descriptions of new species in recent years have made those keys outdated. Thus, our goal in writing this manual was to provide identification keys, descriptions, and distributional notes for amphibians and reptiles of North America, north of Mexico. To our knowledge, it is the first volume to contain identification keys for all species of herpetiles in the U.S. and Canada. Thus, the book should be useful for college level herpetology courses which require students to identify species in any area of the country. While providing the technical details necessary for species identification, we have tried to keep the terminology simple and to give sufficient explanations of terms parenthetically in the text and in the glossary so that interested amateurs and precollege students will also find the book useful. This book makes an aid to identification of amphibians and reptiles available at a more economical price than comprehensive field guides and within a single cover.

The book is organized in the traditional format of the *How to Know* series with identification of species for a particular family treated in a continuous key. Keys to salamanders are followed by keys for anurans, turtles, lizards, snakes, and crocodiles with introductory keys leading to each of the major groups so that the reader can begin at the beginning of the keys or proceed to a particular family depending on level of prior knowledge. We have also provided a key to larval amphibians at the end of the section on amphibians. In addition to the short description of each species in the key, we have given brief descriptions of general distribution (maps in many cases). Natural history notes are given on each family, as well as larger groups. We have provided introductory remarks on characteristics of amphibians and reptiles, references, how to study, capture and keep species, and special notes on snake bite care in the first part of the book. An outline of the classification system we follow, including a list of all higher categories and lists of genera within families, is given at the end of the book. We have included sufficient line drawings to illustrate technical characteristics, as well as representative drawings of most genera and distinctive species.

We followed the suggestions of the committee report ("Standard Common and Current Scientific Names for North American Amphibians and Reptiles", 1978) by Collins et al. of the

Society for the Study of Amphibians and Reptiles in matters of common names. Species described up to January 1982 have been included. Data for constructing the identification keys has come from our own experience, as well as a wide variety of publications and technical papers. Due to geographic variation in diagnostic characters there will undoubtedly be problems with some keys for specimens from some regions of the country. We hope readers will call these problem areas to our attention, so that improvements can be made in future editions.

We appreciate the courtesy and assistance of the editors of this series, particularly Drs. Ed Cawley and John Bamrick. The manuscript was typed by Lola Beach and Kathy Beatty, whose assistance we appreciate. Most of the drawings were done by Emily Oseas to whom we are deeply indebted and extremely grateful.

Royce E. Ballinger
John D. Lynch

School of Life Sciences
University of Nebraska
Lincoln, Nebraska

Introduction

THE SCIENCE OF HERPETOLOGY

Herpetology is the scientific study of amphibians and reptiles. *Herpes* was the Greek word meaning to creep or crawl and a *herpeton* was a creature which crawled. The early Greek scientists made no distinction between reptiles and amphibians although modern scientists consider these groups to represent two distinct classes of chordates.[1] Like other chordates, amphibians and reptiles have a dorsal hollow nerve cord, pharyngeal gill slits at least during early developmental stages, a postanal tail, and a notochord which is replaced by vertebral elements of bone that form the principal supporting axial skeleton as in all animals of the subphylum Vertebrata.

As in other vertebrates there is pronounced cephalization with an anterior brain (surrounded by a protective cranium) which also bears the primary specialized sensory organs. Other than these basic vertebrate characteristics, the similarities between reptiles and amphibians are few, except that both groups are "cold-blooded." Thus, they tend to be distributed in warmer regions of the world, although salamanders have a greater number of species in the north temperate regions. Nevertheless, scientists specializing in the study of amphibians and reptiles often work with both groups and thus the science of herpetology retains the unity established by the Greeks. Herpetology encompasses the study of any aspect of biology dealing with species of either vertebrate class. Historically, herpetologists studied amphibians and reptiles because of their intrinsic interest in the anatomy, physiology, systematics, ecology, or other aspects of these interesting creatures. However, increasing numbers of herpetologists simply use reptiles or amphibians as convenient study systems to answer specific questions of general biological interest.

There are approximately 3000 species of amphibians and about 6000 species of reptiles living today. These numbers represent only a small number of these once highly successful "lower" vertebrate groups. Amphibians were the dominant land vertebrates during late Paleozoic and early Mesozoic times; they share a common ancestry with dipnoan fishes. (See geological chart) Reptiles share a common ancestry with labyrinthondont amphibians, having originated in the Carboniferous period of the late Paleozoic and becoming the dominant land vertebrates of the Mesozoic (age of the reptiles). Therapsid reptiles gave rise to mammals in the Triassic period and birds arose from saurischian reptiles in the Jurassic period.

[1]Technically the Reptilia is a paraphyletic group (see Wiley, E. O., 1981, Phylogenetics, Wiley Interscience, for an Introduction to the theory of classification and systematics).

ERA	Period	Million Years Before Present							
CENOZOIC	Quaternary	2	APODA	URODELA	SALIENTIA	TESTUDINIA	LACERTILIA	SERPENTES	CROCODILIA
	Tertiary	67							
MESOZOIC	Cretaceous	133							
	Jurassic	180							
	Triassic	228							
PALEOZOIC	Permian	274							
	Carboniferous	350	PRIMITIVE REPTILES						
	Devonian	400	PRIMITIVE AMPHIBIANS						
	Silurian	440							
	Ordovician	500							
	Cambrian	600							

Geological distribution of the major Recent groups of Amphibians and Reptiles

INTRODUCTION TO AMPHIBIANS

Modern amphibians (Lissamphibians), whose evolutionary origins can be traced inexactly back into the late Paleozoic or early Mesozoic, represent only one of three subclasses of amphibians which flourished before the dawn of Mesozoic reptiles. The other two subclasses, labyrinthodont and lepospondyl Amphibia, survive only as fossils in layers of sedimentary rock as reminders of the Age of Amphibians.

Three orders of Lissamphibia are generally recognized—the Caudata (or Urodela)[2] for the salamanders, the Salientia (or Anura) for frogs and toads, and the Gymnophiona (or Apoda) for the superficially worm-like caecilians (Fig. 1).

Amphibians are intermediate between aquatic fishes and the terrestrial reptiles but the intermediacy is a discordant one. Like fishes, amphibians are anamniotes lacking the modifications of the early embryo with its extraembryonic membranes (amniotic membranes) and lacking the impervious integument of reptiles (and their descendants). In general, one describes amphibians as having external fertilization and aquatic larvae but, in keeping with the mysteries of their long evolutionary history, some Lissamphibia parallel amniotes with internal fertilization and in lacking aquatic larvae.

Unlike reptiles, amphibians employ the skin as a respiratory device and accordingly have moist (and slimy) integuments unadorned with scales. The integument is abundantly supplied with

[2]Latin ordinal or subordinal names are used herein in preference to Greek names when possible.

glands providing some measure of defense but also aiding the retention of water during the animal's brief forays away from ready access to free water. Without the epidermal scales of reptiles or the dermal scales of fishes, the skin of amphibians constrains them to a life near freshwater. For the most part the xeric environments of deserts and grasslands are left to the more adventuresome reptiles while amphibians creep about in dark, cool, and wet subdivisions of the biosphere. Within these limits, amphibians are broadly distributed. Subterranean, terrestrial, aquatic, and arboreal adaptive zones have been effectively exploited by both frogs and salamanders. Caecilians have not discovered an avenue into the arboreal zone.

Caecilians are essentially tropical in their distribution, and none of the approximately 160 species occurs in temperate North America. *Dermophis mexicanus* ranges north to Veracruz, Mexico. The five currently recognized families of completely limbless amphibians range across the neotropics, subsaharan Africa, the Seychelles Islands, peninsular India, and southwestern Asia onto the islands of the IndoAustralian archipelago. All exhibit internal fertilization and most retain the young in the oviducts of the female until metamorphosis. Caecilians have poorly developed eyes (sometimes covered over with bone) and use a pair of retractable tentacles located between the eye and nostril as sensory devices. Most lack a tail but the tailed caecilians (Ichthyophiidae and Rhinatrematidae) have short but evident tails.

Figure 1   Orders of Amphibia

Salamanders and frogs (including toads) are abundant although secretive organisms found in north temperate regions. Both groups invade the tropics but frogs have done so in a major way whereas among salamanders only the most advanced group of the lungless salamanders (Plethodontidae) has done so and only into the Neotropics.

Eight of the nine living families of salamanders occur in temperate North America. Only the Asiatic family Hynobiidae is not represented in the fauna of the United States. One hundred and twelve of the approximately 330 known species of salamanders occur natively in the United States. In contrast, eight of the 19 living families of frogs and toads occur natively in temperate North America. The families Brachycephalidae, Centrolenidae, Dendrobatidae, Pseudidae, and Rhinodermatidae occur only in the Neotropics. The family Discoglossidae is Eurasian. The Old World leptodactyloids (African Heleophrynidae and Australian Myobatrachidae), the Old World tree frogs (Hyperoliidae and Rhacophoridae), and the aquatic Pipidae are not native to the temperate North America. Nevertheless, human intervention has resulted in the African Clawed Frog (*Xenopus laevis,* Pipidae) establishing itself in some stream systems in southern California. Only 80 of the more than 2900 known species of frogs and toads are native to temperate North America.

Modern Amphibians share a suite of characteristics which tell us more about what they are not than what they are. A variety of features can be cited to compel recognition that this is a tetrapod group but the other traits are cited to show that these organisms are not reptiles (or their descendants). The orders are dramatically separated from one another. Caecilians are limbless, have either very short tails or no tails, are elongate, slender organisms, have tentacles and insignificant eyes, have a complex arrangement of bones in a strengthened skull to aid presumably in burrowing, and have an intromittant organ in males (and internal fertilization). In contrast, frogs have no tail, have modified the pelvis in designing a jumping device, have very abbreviated vertebral columns, fusion of the bones in the zeugopodium (lower limb), elongated hind limbs (with elongate tarsi), a fragile skull, usually have a sternum, have no teeth on the lower jaws (excepting a peculiar tree frog found in Ecuador) and most exhibit external fertilization. Salamanders have long, powerful tails, an ordinary amphibian pelvis (short ilia), an ordinary vertebral column (approximately 20 trunk vertebrae, except in the attenuate species), independent bones in the lower limb, approximately equal-sized fore and hind limbs (except in sirenids), a fragile skull, and like caecilians have teeth on the upper and lower jaws. Salamanders exhibit both external and internal fertilization but the vast majority exhibit internal fertilization. Internal fertilization is accomplished by the male depositing a spermatophore (a capsule containing sperm) on the substrate. The female picks the spermatophore up with her cloacal lips and sperm move to spermathecae (tubules located on the dorsal surface of the cloaca).

North American students are exposed to a very limited variation in amphibian life histories. Some of our plethodontid salamanders and most of our leptodactylid frogs lay large, terrestrial eggs and the larvae complete development and metamorphose within the confines of the egg capsule. The vast majority of our amphibians exhibit a monotonous life cycle of (a) eggs deposited in water, (b) hatching into a free-swimming larva, (c) metamorphosis into a 'terrestrial' juvenile,

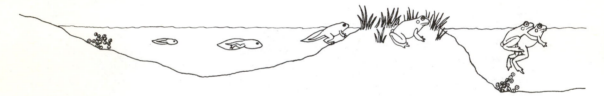

Figure 2   Simple life history of a frog

(d) growth and maturation, and (e) a return to water to breed (Fig. 2). The student should bear in mind that exotic amphibians, chiefly frogs, have explored many reproductive modes. Fully 25 percent of all frog species use direct development (no tadpole). Even less pedestrian reproductive modes occur. The Australian frog *Rheobatrachus silus* swallows its young which develop in their mother's stomach! *Rhinoderma* of Chile takes the tadpoles into the mouth where they develop in the male's vocal sacs. Several frogs utilize pouches on the back *(Gastrotheca)* or side *(Assus)*. The African toads *Nectophrynoides* and a species of Puerto Rican *Eleutherodactylus* give live birth to froglets which develop in their mother's oviducts.

In spite of the superficial similarity in the modal reproductive strategy of both frogs and salamanders, the larvae are easily separated on the basis of the number of pairs of gills, mouth parts, and sequence of limb development.

Metamorphosis in frogs signals a major reorganization (Fig. 3). The tail is lost, an adult mouth replaces the larval mouth, the tympanum becomes obvious, the tongue appears, the eyes enlarge, and internally the long coiled herbivorous gut is replaced by a short carnivorous gut (Fig. 4).

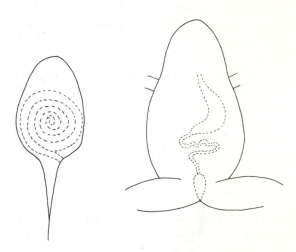

Figure 4   Tadpole and adult frog digestive tracts

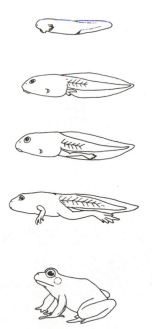

Figure 3   Developmental stages of a frog

Salamanders present a special problem in that metamorphosis is subtle. Obvious changes include acquisition of eyelids, loss of external gills, tail and body fins, and gill slits. Other changes are frequently in acquisition of a color pattern. No substantive changes in proportions or body form occur. The larval arrangement of dentition changes in many salamanders so that the vomerine teeth cease to parallel those of the maxilla and premaxilla (Fig. 5).

Many salamanders exhibit a phenomenon called paedogenesis where although maturing reproductively they remain anatomically immature retaining the gross morphology of larvae. Sirenids are perhaps the most dramatic. Not only do they fail to develop maxillae, they do not acquire hind limbs. Proteids and some plethodontids acquire hind limbs but retain the early larval skull morphology (no maxillae). Cryptobranchids and Amphiumids loose fins and gills and acquire maxillae but retain gill slits. Some ambystomatids and dicamptodontids are also pae-

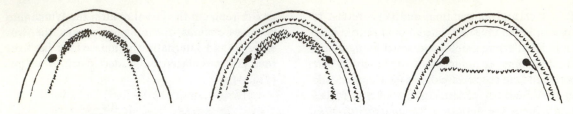

Figure 5   Mouths of larval (left) and adult (right) salamanders

dogenetic. *Ambystoma tigrinum* is a facultative paedogene in that metamorphosis is variable within and between populations. Those salamander species or populations in which reproductive maturity is achieved without complete metamorphosis are termed neotenic or paedogenetic.

INTRODUCTION TO REPTILES

Modern reptiles represent a small fraction of those present in Mesozoic times. Of 16 orders of reptiles only four have living representatives which include three of the six described subclasses. Living orders (Fig. 6) of reptiles include Testudinia (or Chelonia) for the turtles, Squamata for the lizards and snakes, Crocodylia for the alligators and crocodiles, and Rhynchocephalia for a single species, the tuatara *(Sphenodon punctatus),* of New Zealand. Except for the latter, members of the other orders are found world wide but are most common in tropical and xeric habitats.

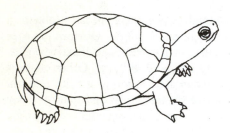

Figure 6   Orders of Reptiles

North American turtles represent most living families of the suborder Cryptodira (S-necked turtles) including the Chelydridae (snapping turtles), the Emydidae (pond turtles and relatives broadly distributed throughout Central America, into South America, and from Europe to the East Indies), the Testudinidae (land tortoises also found in South America, Africa, and from the Mediterranean to Southeast Asia), the Kinosternidae (musk and mud turtles also found in Central and South America), the Trionychidae (soft-shelled turtles also distributed across Central Africa and from the Middle-East across India to Southeast Asia and the East Indies and up the east Asian coast beyond and including Japan), and the sea turtle families (Cheloniidae and Dermochelyidae). Three extant cryptodire families not represented include the Dermatemydidae (one species, *Dermatemys mawi,* in Central America), the Platysternidae (one species, *Platysternon macrocephalus,* in southeast Asia), and the Carettochelyidae (one species, *Carettochelys insculpta,* of northern Australia and southern New Guinea). The side-necked turtles (suborder Pleurodira) including the Pelomedusidae (found in South America and Africa) and the Chelidae (found in South America, Australia and New Guinea) are not represented in North America.

The squamate reptiles (lizards and snakes) are represented in North America (north of Mexico) by nine families of lizards (Rhineuridae, Anguidae, Anniellidae, Xantusiidae, Gekkonidae, Teiidae, Scincidae, Helodermatidae, and Iguanidae) and five families of snakes (Leptotyphlopidae, Viperidae, Micruridae, Colubridae, and Boidae). Additional lizard and snake families not found in North America (north of Mexico) include lizard families Pygopodidae (snake- or worm-like lizards related to the geckos), Agamidae (agamids, iguanid counterparts in the old world and Australia), Chamaeleonidae (true chamaeleons), Xenosauridae (*Xenosaurus* of Central America and *Shinisaurus* of southern China), Lanthanotidae (*Lanthanotus borneensis* of Borneo), Dibamidae (the limbless *Dibamus* of the East Indies and *Anelytropsis* of eastern Mexico), Varanidae (monitor lizards), Cordylidae (plated lizards of Africa), and Lacertidae (lacertids, teiid counterparts in the old world). The amphisbaenians (ring-lizards) include in addition to the Rhineuridae, the Mexican Bipedidae, the Trogonophidae (north Africa and Socotra Island) and the Amphisbaenidae (southern Europe, adjacent Africa, sub-Saharan Africa, the Greater Antilles and warm lowland South America).

Snake families not found in North America include the Anomalepidae (blind, burrowing snakes of southern Central America and South America), Typhlopidae (blind snakes found in the tropics world wide), Acrochordidae (wart snakes of southeast Asia, the East Indies and northern Australia), Aniliidae (*Anilius scytale* in tropical America), Uropeltidae (rough-tailed snakes of eastern India and southeast Asia), Loxocemidae (*Loxocemus bicolor* of Mexico and Central America), and Elapidae (cobras and allies of Africa, southeast Asia, East Indies, and especially Australia). Some authors also recognize the sea snakes (Hydrophiidae, not represented in North America north of Mexico) as one or two distinct families. The familial systematics of snakes is highly controversial at this time. Surveys are incomplete but the large Colubridae family may actually consist of at least a dozen distinct families.

Of the living Crocodylia, the Crocodylidae are represented in North America by *Alligator* and *Caiman* genera (subfamily Alligatorinae) and *Crocodylus* (subfamily Crocodylinae). The other subfamily (Tomistominae) represented by the false gavial, *Tomistoma schlegeli,* is found on the Malay Peninsula, Sumatra, and Borneo. Likewise, the family Gavialidae (*Gavialis gangeticus* restricted to India and Burma) is not represented in North America.

Reptiles have characteristics generally intermediate between amphibians and those of mammals and birds. Reptiles evolved characteristics for a total independence from water unlike amphibians which generally must return to water to breed. The complete terrestrial existence of reptiles was made possible with the evolution of the amniote egg as well as an integument of epidermal scales and improved respiratory, excretory, and circulatory mechanisms.

The amniote egg (Fig. 7), found also in birds and mammals, consists of a series of membranes surrounding the developing embryo. The embryo develops within the fluid of the amniotic cavity and thus is protected from desiccation in the terrestrial environment. The chorion membrane surrounds the entire embryo and gives rise to an outer protective shell. The shell may be leathery or parchment-like such as in lizards, some turtles, and snakes which must bury their eggs in moist ground or may contain calcium and appear almost bird-like as in some turtles, crocodiles, and a few lizards.

The amniote egg has two other membranes. The yolk sac contains energy rich yolk used by the developing embryo as a source of nutriment as well as varying amounts of albumen ("egg white"). Albumen is a gelatin substance serving as a source of water. Amounts of albumen vary depending on the nature of the shell and its environment. Embryos of reptiles which lay eggs in moist ground, obtain much of their water through the shell thus there is less albumen. In general, reptilian eggs contain much less albumen than do eggs of birds. The final membrane (allantois) serves as an excretory depository. The primary excretory product of reptiles is uric acid, a non-water soluble, non-toxic substance which requires no water for its disposal. This metabolic adaptation further aids reptiles to conserve water in contrast to fishes and amphibians which produce primarily ammonia or urea, both of which are toxic, water soluble, and require a constant supply of water to flush the excretory products from the body.

The outer skin of reptiles consists of a horny layer of epidermal scales. This keratinized layer provides a seal against water loss, a condition not possible in amphibians in which the integument plays an important role in respiration. The outer epidermal layer is periodically shed or worn off and replaced by scales developed from underlying layers. The number, size, shape, and/or arrangement of epidermal scales are usually species-specific and thus provide excellent characters for identification.

With the loss of the integument as a respiratory organ (compared to amphibians), an increased efficiency of gas exchange was needed. This has been accomplished with modifications in both the respiratory and circulatory systems. The lungs of reptiles are larger and more subdivided than are those of amphibians. In addition, the ribs play an active role in ventilation (except in turtles) rather than there being total dependence on buccopharyngeal movements as in amphibians. In turtles the ribs are not movable and thus breathing is accomplished by contraction of specialized trunk muscles against certain coe-

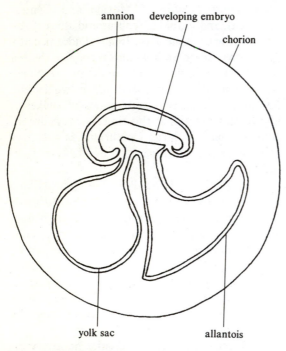

amnion   developing embryo

chorion

yolk sac   allantois

Figure 7   Amniote egg

lomic membranes. There is also a partial (complete in Crocodilians) separation of the air passage from the mouth by a secondary palate in the roof of the mouth. The palate allows for feeding without interference with breathing.

The terrestrial habit of reptiles also necessitated alterations in the circulatory system, particularly improvements in circulation to the lungs. The heart of crocodilians contains four chambers like the hearts of birds and mammals and there is a complete separation of the oxygenated and deoxygenated blood. In other reptiles the atria are separated with the right atrium receiving deoxygenated blood from the body and the left atrium receiving oxygenated blood from the lungs, but the ventricle is only partially separated by an incomplete interventricular septum. In spite of this the rhythm of the heart beat allows only partial mixing of oxygenated and deoxygenated blood before it is pumped to the body.

Reptiles also have a more efficient kidney of the metanephric type found also in birds and mammals. The opisthonephric kidney typical of amphibians is found in early developmental stages in reptiles but is replaced by the more compact and efficient metanephros by the time that birth or hatching occurs. Although the reptilian metanephros does not contain Loops of Henle to conserve water as occurs in birds and mammals, the increasing secretion of uric acid by reptiles in dryer habitats provides a mechanism to prevent excess water lost due to excretion.

Thus, adaptations such as the amniote egg and improved respiratory, circulatory, and excretory systems allow reptiles to lead a completely terrestrial life. Although some reptiles (crocodiles and certain turtles) are found in aquatic habitats like amphibians, reptiles generally occur in habitats not available to amphibians.

HOW TO STUDY HERPETILES

It is possible to begin study of amphibians and reptiles by visiting the local library (see next section on general references) or local zoo, but firsthand knowledge gained from observing specimens in nature is preferable. Some species may be observed easily (for example, diurnal lizards or turtles) as they attend to their daily activities whereas other species, amphibians in general and snakes in particular, are usually more secretive in their habits. Thus, it is sometimes of interest to capture specimens for close examination or to keep them in captivity to observe their behavior.

Methods of Capturing Herpetiles

Most frogs and toads, salamanders, lizards, snakes, and some turtles are most conveniently captured by hand as one wanders through suitable habitat in search of specimens. Many of our aquatic turtles are more difficult to capture and require special techniques. These turtles and aquatic amphibians are sometimes taken in seines. Juvenile specimens sometimes can be captured with a dipnet. In some aquatic situations, traps are the only effective method of capture. Basking traps and hoop or Fike nets (Fig. 8) function to capture these more elusive turtles. Before employing traps or nets, check with the appropriate state agency (Fish and Game Departments) concerning regulations.

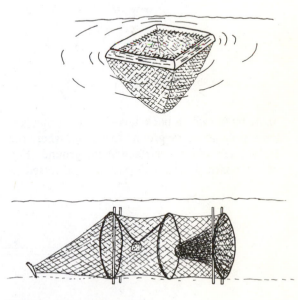

Figure 8   Turtle traps

Many lizards can be captured using nooses. A noose is made from strong string (carpet thread, dental floss, braided fishing line all work well, monofilament fishing line is too stiff) by tying a slipknot or eyeknot then running the line through the eye to allow the noose to close easily (Fig. 9a).

Figure 9a   Lizard noose

Putting the noose on the end of a cane pole, fishing rod or straight stick works well as an extension device. For some nocturnal animals or wary lizards can-traps are employed. A can-trap is

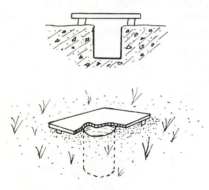

Figure 9b   Pit-fall traps

made by burying a bucket, coffee can or one gallon plastic jar in the ground to a depth where the top is flush with the surface of the ground (Fig. 9b). A board or other suitable material raised on

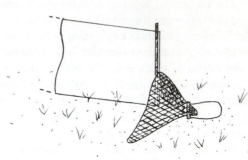

Figure 9c   Drift fence with funnel trap

pegs should be placed across the top of the can to provide shade and prevent the animal from overheating during the day. Drift-fences (Fig. 9c) of hardware cloth or sheetmetal can also be effective in catching reptiles or amphibians. Place a container with a funnel made of hardware cloth at each end of the fence to collect animals traveling along the fence.

A sturdy snake-hook (Fig. 10) can be used for overturning rocks or boards and for moving debris without exposing fingers to a venomous snake or lizard. A snake-hook is a useful device for capturing many snakes. Once the hook is pressed across the snake's head, one can grasp the head using three fingers. This grip protects you

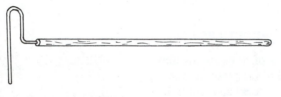
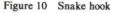

Figure 10   Snake hook

from a bite. This capture method is among the safest for venomous snakes but these animals pose a serious and continuing hazard to those who keep them. Tongs and sturdy nooses are also of use in handling venomous snakes.

Care of Captives

Once captured, you may wish to simply examine your captive closely recording whatever features you wanted to see. If this was your goal, release the animal where you caught it. If you want to observe your captive for longer periods some sort of cage is required.

Transporting captives from the field to home requires a few supplies. Reptiles can be transported in cloth bags but amphibians require protection from desiccation. Plastic jars supplied with wet paper towels work well (Fig. 11). Both amphibians and reptiles need protection from exposure to direct sun.

Aquaria with screen lids serve as suitable cages for housing amphibians or reptiles. A water supply and a place of concealment should be provided as minimal furnishings. Most specimens also need a place to dry-off or to sun (frequently atop the concealment site). More elaborate cages (Fig. 11) may be preferred and are usually beneficial for longer term observations but the minimal furnishings should never be excluded. Larval amphibians can be kept in an aquarium half-filled with water.

If you are keeping your captive for more than a few days, you will need to provide food. All frogs, salamanders, and snakes are carnivores requiring live prey as are most lizards. In general, frogs, salamanders, and lizards will feed on insects (e.g., crickets, mealworms, etc.). Most amphibians readily accept worms. Food should be offered at least every other day (daily is best). Larval frogs and toads (tadpoles) are usually herbivores. Boiled lettuce, spinach, or similar materials will provide food for most. Larval salamanders are carnivores. Small aquatic organisms are needed as food for these animals. Some

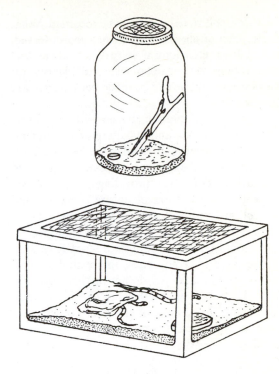

Figure 11   Cages for live specimens

herbivorous lizards (chuckwallas, desert iguanas) prefer fruits and flowers. Many turtles accept earthworms, small fish, or pieces of meat (beef-heart is an excellent food) but others prefer plant materials. Many common snakes eat earthworms (small ground dwelling snakes) or fish (water snakes) but many others eat mice (bull, fox, racers, rat snakes) as do the terrestrial vipers. Hognosed snakes primarily eat toads. Snakes feed less often than do other amphibians and reptiles but should be offered prey once a week. In feeding your captives, consult the references at the end of the chapter for additional details of food preferences for particular species.

When you have tired of your captive or wish to secure a different one, release the captive where it was captured.

Keep a notebook in which to record your observations. Your notebook serves to provide reference concerning where you found your specimen and what it seemed to be doing. During captivity, additional notes can be recorded. What it ate, how often, what behaviors you observed, its growth, etc. So long as you are actively adding observations to your notebook, a justification for keeping the captive can be made because you are learning.

A word of caution. Faced with increasing pressure from man and his altered environment, many amphibians and reptiles are losing a critical battle—for living space. Conservationists and legislative bodies are providing much needed protection in the form of nature preserves and protection from molestation or capture. Those amphibians and reptiles now under legal protection should be left alone. Learn what laws exist in your area to protect these interesting animals and be certain that you do not further endanger animal populations. Methods of capture may be regulated. The number of specimens in captivity may be regulated. Do not become part of the problem!

Snake Bite First Aid

Venomous snakes should be handled with extreme caution and care should be exercised when working in the field in areas where venomous snakes are common. A bite from a venomous snake is a serious matter and should be treated as a medical emergency. You should familiarize yourself with the snakes in your area so that positive identification can be made in the case of a bite. It is good practice to carry a snake bite kit in the field with you. If a bite occurs from a venomous snake, immediate steps should be taken to obtain professional medical assistance. If it is certain that assistance cannot be obtained within 30 minutes, then first aid should be administered.

Keep the victim calm in a sitting or reclined position. Remove rings and bracelets and tie a constricting band around the arm or leg between the bite and body. The band should be tight enough to impede flow of superficial blood and lymph but not so tight as to constrict deep flow. The band should not be so tight that you cannot force your finger under it. If swelling occurs, the band may need to be loosened or moved upward. A sharp knife or razor blade should be sterilized with a flame or antiseptic if either are available. Even if sterilization is not available, make a cut through each puncture wound of the bite parallel to the axis of the arm or leg. Do not make "X" marks or cut across the axis. The cut should be no more than 1/8 inch deep and 1/4 inch long. Apply a suction cup to the cut for 3 to 5 minutes and then empty the cup. Continue this procedure for at least an hour while on the way to a hospital or doctor's office. If a suction cup is ot available you can use your mouth provided you have no sores or cuts in it which would permit the poison to enter your blood system.

Antivenin is useful in treating snake bites but should only be administered by a doctor. Antivenin may not always be available at the nearest medical facility so if you are on a long trip or have a remote campsite or summer home it is useful to have antivenin which you can take to the hospital with you. Antivenin can only be obtained by prescription in cooperation with your personal physician. Because venom differs between species the antivenin available by prescription is usually for American Pit Vipers. A doctor can obtain antivenin for coral snake bites by contacting the Center for Disease Control, U.S. Public Health Service, Atlanta, Georgia. In the event of a bite by an exotic venomous snake (such as those held in zoos or as personal pets, a very dangerous practice) the doctor can obtain information on the availability of antivenins for foreign species by contacting the Antivenin Index Center of the American Association of Zoological Parks and Aquariums. This center has a 24 hour answering service at the Oklahoma Poison Control Center of the Oklahoma City Zoo.

Remember, the best treatment for snake bite is prevention. Do not handle venomous snakes unless you have had considerable experience with non-venomous species first. Handle "dead snakes" with extreme caution. Reflex actions can cause a dead snake to bite and the venom remains potent for a considerable length of time after the snake has died.

General References

The following books are recommended as additional sources of material on amphibians and reptiles. References cited in these books provide further information for the more serious reader.

Bellairs, A. 1969. The Life of Reptiles. Weidenfeld and Nicholson, London, 2 volumes.

Blair, W. F., A. P. Blair, P. Brodkorb, F. R. Cagle, and G. A. Moore. 1968. Vertebrates of the United States. McGraw Hill, New York.

Bishop, S. C. 1947. Handbook of Salamanders. Comstock, Ithaca.

Carr, A. 1952. Handbook of Turtles. Cornell Univ. Press, Ithaca.

————. 1963. The Reptiles. Time Inc., New York.

Cochran, D. M. 1961. Living Amphibians of the World. Doubleday, Garden City.

Conant, R. 1975. A Field Guide to Reptiles and Amphibians of Eastern and Central North America. Houghton Mifflin Co., Boston, Second Edition.

Ernst, C. H. and R. W. Barbour. 1973. Turtles of the United States. U. Press of Kentucky, Lexington.

Goin, C. J., O. B. Goin, and G. R. Zug. 1978. Introduction to Herpetology. Freeman and Co., San Francisco.

Harrison, H. H. 1971. The World of the Snake. Lippincott, New York.

Leviton, A. E. 1972. Reptiles and Amphibians of North America. Doubleday, New York.

Minton, S. A. and M. R. Minton. 1973. Giant Reptiles. Scribner's, New York.

Noble, G. K. 1954. The Biology of the Amphibia. Dover, Garden City.

Oliver, J. A. 1955. The Natural History of North American Amphibians and Reptiles. Van Nostrand, Princeton.

Peters, J. A. 1964. Dictionary of Herpetology. Hafner, New York.

Pope, C. H. 1955. The Reptile World. Knopf, New York.

Porter, K. R. 1972. Herpetology. Saunders, Philadelphia.

Schmidt, K. P. and R. F. Inger. 1957. Living Reptiles of the World. Doubleday, New York.

Smith, H. M. 1946. Handbook of Lizards. Comstock, Ithaca.

Stebbins, R. C. 1966. A Field Guide to Western Reptiles and Amphibians. Houghton Mifflin, Boston.

Wright, A. H. and A. A. Wright. 1949. Handbook of Frogs and Toads of the United States and Canada, Third Edition. Comstock, Ithaca.

———— and ————. 1957. Handbook of Snakes. Comstock, Ithaca, 3 volumes.

How to Use the Keys

The remainder of this book is designed primarily for the identification of the species of amphibians and reptiles but brief introductions are given for each major group (order or suborder) and family. If it is evident that the specimen you wish to identify is a salamander, frog or toad, turtle, lizard, snake or crocodile it is possible to proceed directly to the key to the famlies of these groups. If you are uncertain then you should begin with the key to the orders of amphibians (p. 15) or orders and suborders of reptiles (p. 94) or on page 15 with the distinction between amphibians and reptiles.

The identification of a particular specimen is accomplished by use of a simple identification key. The statements in these keys occur in pairs denoted by "a" and "b". Each pair or "couplet" is numbered and the two parts of a particular number comprise a set of contrasting statements about the specimen you wish to identify. By comparing "a" and "b" with the specimen one is able to determine which statement applies, then out at the right is a number indicating which couplet should be considered next. This process of choosing between contrasting statements should be repeated until the name and description of the specimen appears at the right of a couplet statement. If the key has been followed correctly the description should agree with the specimen being considered. If it does not the process should be carefully repeated to discover where the error was made.

Distinction between Amphibians and Reptiles

1a Skin with covering of epidermal scales (see page 94 for key to orders and suborders of reptiles, see fig. 6 for body forms)............. .... **Reptilia**

1b Skin without covering of epidermal scales (see page 15 for key to orders of amphibians, see fig. 1 for body forms)................... .... **Amphibia**

Key to Orders of Amphibians

1a Tail long and conspicuous (salamanders) ....(p. 17) **Caudata**

1b Tail absent or short and inconspicuous (frogs and toads)...................................... .... (p. 55) **Salientia**

SALAMANDERS,
Caudata (Urodela)

Salamanders vary considerably in proportions from slender elongate organisms *(Amphiuma)* to relatively stocky organisms *(Ambystoma, Desmognathus)*. Most have robust limbs but those of amphiumids are spindle-like and minute. Most are terete (round in cross-section) at midbody but *Cryptobranchus* is prominently depressed (broad flat head and body). Most salamanders have long tails (tail length $\geq$ standard length) but in others the tail is only half as long as the standard length. Standard length (SL) is defined as the distance from tip of snout to posterior angle of vent. In some plethodontid salamanders the tail is constricted immediately posterior to the vent. The anal opening of salamanders is an elongate slit and is swollen in sexually mature individuals. Typical structural features of salamanders are shown in Fig. 12.

The variable expression of paedogenesis (permanent retention of larval characteristics after reproductive maturation) provides many features of value in distinguishing salamanders. The hind limbs are absent in sirenids. Sirenids, proteids, and some ambystomatids, dicamptodontids, and plethodontids are perennibranchs (retain gills as adults). Others such as amphiumids and cryptobranchids lose the gills but retain gill slits.

Other external traits useful in species discrimination include the presence of cranial crests in *Notophthalmus,* nasolabial grooves in non-paedogenetic Plethodontids, costal grooves (and folds) in many salamanders reflect the number of trunk vertebrae (see Fig. 12), a gular fold across the neck, and the presence of lateral body folds.

Relative limb length is often expressed by adpressing the limbs against the body and recording the separation (or overlap) of toe tips in terms of costal folds (Fig. 13).

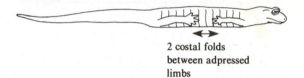

2 costal folds
between adpressed
limbs

Figure 13   Method of counting costal folds, between adpressed limbs

In most salamanders four fingers occur on the forelimbs and five on the hind limb but reductions as dramatic as to only one toe on each limb occur *(Amphiuma pholeter)*. The most common variant is four toes on each limb. Toes vary in terms of length from short digits such as in some plethodontids to the long slender digits of some *Ambystoma* species.

Some features are evident inside the mouth. The arrangements of vomerine teeth are subject to paedogenetic interference but distinctive patterns occur in many salamander families (Fig. 14). The tongue of some plethodontid salamanders is stalked (all edges free). Desmognathine plethodontids are unique in having the lower jaw immobile. They raise the top of the head in order to open the mouth.

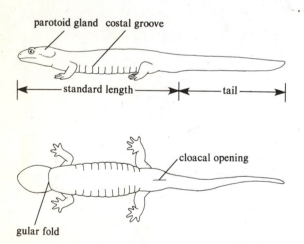

parotoid gland costal groove

standard length —— tail

cloacal opening

gular fold

Figure 12   Typical structures of a salamander

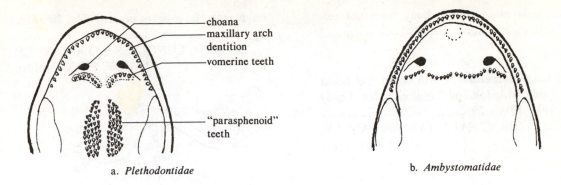

choana
maxillary arch
dentition
vomerine teeth

"parasphenoid"
teeth

a. *Plethodontidae*

b. *Ambystomatidae*

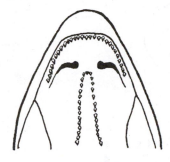

c. *Salamandridae*

Figure 14   Teeth arrangement in salamander families

All plethodontid salamanders lack lungs. Reduced lungs also occur in the ambystomatid *Rhyacotriton*.

Most salamanders have obviously wet and slimy skins but roughened skins are found at various stages of life in both genera of salamandrids.

KEY TO FAMILIES OF SALAMANDERS

1a One pair of legs present; elongate, eel-like salamanders with gills (Fig. 15)................ .... (p. 23) SIREN FAMILY, Sirenidae

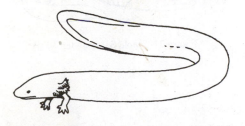

Figure 15   Siren

1b Both fore and hind limbs present; body shape variable 2

2a Forelimbs and hind limbs minute; body elongate, eel-like salamanders with a long snout and one pair of gill slits (Fig. 16).... .... (p. 28) CONGO EEL FAMILY, ... Amphiumidae

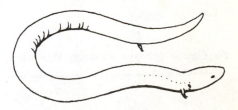

Figure 16   Congo eel

2b Limbs more robust; body form not eel-like 3

3a Large salamanders with broad, flat heads and having folds of skin along sides of body (Fig. 17) (p. 29) ....GIANT SALAMANDER FAMILY, Cryptobranchidae

Figure 17   Giant Salamander

3b Size minute to large but head not greatly flattened; no skin fold on flanks 4

4a Salamanders bearing 2–3 pairs of gills.. 7

4b Gills not present....................................... 5

5a Costal grooves not apparent; skin rough or smooth (Fig. 18).. .... (p. 31) NEWT FAMILY, Salamandridae

Figure 18   Newt

5b Costal grooves distinct; skin slimy and smooth 6

6a Nasolabial grooves present (Fig. 20) (Fig. 19)... ..... (p. 34) LUNGLESS SALAMANDER FAMILY, Plethodontidae

Figure 19   Lungless Salamander

Figure 20   Nasolabial groove

6b No groove between nostril and lip (Fig. 21).. (p. 20) MOLE SALAMANDER FAMI-LIES, Ambystomatidae, Dicamptodontidae

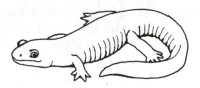

Figure 21   Mole Salamander

7a Fin on upper surface of tail extends onto body (Fig. 22).. .... (p. 90) **Larval and some Neotenic Salamanders**

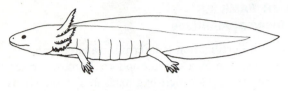

Figure 22

Figure 22   Larval salamander with tail fin

7b Fin on upper side of tail not extending more craniad than to level of hind legs (Neotenic Salamanders) (Fig. 23)............................ 8

Figure 23   Neotenic Salamander

8a Gills bushy; four toes on hind feet; tail length contained two or more times in standard length (Fig. 24) .... (p. 29) **MUDPUPPY SALAMANDER FAMILY, Proteidae**

Figure 24   Mudpuppy

8b Gills slender (short or long); five toes on hind feet; tail length contained no more than 1.5 times in standard length (usually about 1.0 times) 9

9a No costal grooves evident .... (p. 31) **Salamandridae**

9b Costal grooves evident 10

10a Body pigmentless (or reduced to a fine network of melanophores) (Fig. 25)................ .... (p. 34) **Plethodontidae**

Figure 25   Troglodytic Salamander

10b Body with pigment................................. 11

11a Body stout (depth of body contained less than 3 times in distance between limbs) .... (p. 20) Ambystomatidae, (p. 20) Dicamptodontidae

11b Body slender (depth of body contained 3–8 (usually 5–8) times in distance between limbs) .... (p. 34) Plethodontidae

MOLE SALAMANDER FAMILIES
Ambystomatidae and Dicamptodontidae

Although recently separated, these two families are traditionally confused (Fig. 26). The only living dicamptodontids are the two species of *Dicamptodon* whereas the living ambystomatids include 26 species of *Ambystoma,* 3–4 species of the Mexican genus *Rhyacosiredon,* and the monotypic *Rhyacotriton.* Mole salamanders are generally large animals, ranging in size from some of the diminutive eastern forms (8 cm. total length) to large examples of the Tiger salamander (33 cm. total length).

Mole salamanders are distributed from extreme southeastern Alaska, James Bay, and Labrador south to the southern edge of the Mexican Plateau. These are frequently burrowing animals and thus not normally encountered. Most encounters are products of spring and fall migrations.

These salamanders superficially resemble lungless salamanders and Mudpuppies. Lungless salamanders have nasolabial grooves and Mudpuppies have only four toes on the hind foot (never five toes).

Neoteny is common in these families. Some species are permanently neotenic (e.g., *Rhyacosiredon*), others rarely metamorphose (e.g., *Dicamptodon*), and yet others display facultative neoteny (e.g., *Ambystoma tigrinum*). Neotenic individuals are often rather large (20–25 cm. total length) and lack the color patterns of metamorphosed individuals. These neotenic individuals are popularly termed axolotls.

Fertilization is internal, transfer being accomplished by a spermatophore. In most species, females deposit their egg mass in shallow ponds and then leave the developing embryos. In autumn, the marbled salamander deposits her eggs in depressions that will later fill with rain water. Until the eggs are covered with water, she guards her clutch.

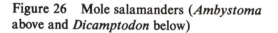

Figure 26

Figure 26 Mole salamanders (*Ambystoma* above and *Dicamptodon* below)

1a Eyes large, eye length equal to or greater than distance from anterior edge of eye to tip of snout (Fig. 27a)...............................
.... Olympic Salamander, *Rhyacotriton olympicus* (Gaige)

a. *Rhyacotriton*

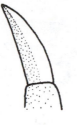

b. *Ambystoma*

Figure 27

Figure 27   Large eye of *Rhyacotriton* (a) compared to other mole salamanders (b)

A small species (44–64 mm SL) with large eyes, square ventlobes (males only), reduced lungs, and no nasal bones. Olive or brown above, yellow to orange below. The Olympic Peninsula of Wash. south to northern Calif.

1b Eyes small, eye length contained within distance between eye and tip of snout (Fig. 27b).. 2

2a Costal grooves indistinct; teeth unicuspid (Fig. 28a), laterally compressed; brown salamanders with black reticulation....... 3

2b Costal grooves distinct; teeth bicuspid (Fig. 28b), not compressed; pattern various, but not brown with black reticulation 4

a   b

Figure 28

Figure 28   Unicuspid (a) and bicuspid (b) teeth

3a O-2 costal folds between tips of adpressed limbs; maturity at 69–104 mm Standard Length............. Cope's Giant Salamander, *Dicamptodon copei* Nussbaum

A small, neotenic species of *Dicamptodon* found on the Olympic Peninsula and south to the Columbia River in Wash.

3b Toe tips of adpressed limbs overlapping, or if separated, separation no more than 0.5 costal folds; maturity usually reached at 120 mm Standard Length......................... .... Pacific Giant Salamander, *Dicamptodon ensatus* (Eschscholtz)

Figure 29

Figure 29 *Dicamptodon ensatus*

Dicampton ensatus frequently metamorphoses. Adults are brown or gray with a network of black reticulations (Fig. 29). Adults range from 85–205 mm SL. Extreme southern British Columbia south to the San Francisco Bay region in Calif. excluding the Olympic Peninsula. Isolated populations in northern Id.

4a Pattern consisting of light spots (as large as or larger than eye) or light transverse bars (as wide as eye, at least laterally), see Fig. 30a...................................... 5

4b Pattern not as above (usually pale flecks or thin transverse lines or vermiculations, sometimes mid-dorsal stripe) see Fig. 30b...................................... 8

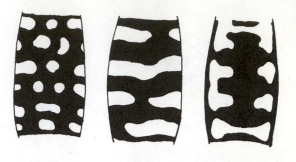

a

b

Figure 30

Figure 30 Color pattern variations in *Ambystoma*

5a Pattern consisting of pale transverse bands.. **6**

5b Pattern consisting of small to large spots.. **7**

6a 2–5 bold light crossbands on head and trunk; 4–7 pale rings on tail; pale markings reach venter (Fig. 31)........................ .... Ringed Salamander, *Ambystoma annulatum* Cope

6b 4–7 pale crossbands on back, not reaching venter (Fig. 32)............................... .... Marbled Salamander, *Ambystoma opacum* (Gravenhorst)

Figure 32

Figure 31

Figure 31   *Ambystoma annulatum*

Color is dark brown to black with light tan to yellow or whitish rings; tongue bearing median furrow (subgenus *Linguaelapsus*); 14–23 cm total length; central Mo. to western Ark. and eastern Okla.

Figure 32   *Ambystoma opacum*

Black with white or gray markings, sometimes forming double dorsal stripe or ladder shape; stocky, small (9–12 cm total length) species of the subgenus *Linguaelapsus*. Distribution (Fig. 33).

Figure 33

Figure 33   Distribution of *A. opacum*

7a A row of pale spots along dorsolateral area of head, body and tail.. Spotted Salamander, *Ambystoma maculatum* (Shaw)

Figure 34

Figure 34   Distribution of *A. maculatum*

Stout-bodied; pale spots yellow or orange in life; ground color blue-black; belly slate-gray; 15–25 cm total length. Distribution (Fig. 34).

7b Pale spots distributed over entire dorsum, not forming two rows (Fig. 35).. Tiger Salamander, *Ambystoma tigrinum* (Green)

Figure 35

Figure 35   *Ambystoma tigrinum*

Dorsum dull black to dark brown; belly olive to yellow and black; dorsal spotting variable (Fig. 36), spots yellow to olive; 15–33 cm total length. Distribution (Fig. 37).

Figure 36

Figure 36   Color pattern variations in *A. tigrinum*

Figure 37

Figure 37   Distribution of *A. tigrinum*

8a Tongue with a median furrow (Fig. 38a); folds or ridges diverge from furrow (subgenus *Linguaelapsus*)........................ 9

8b Tongue lacking median furrow (Fig. 38b) 11

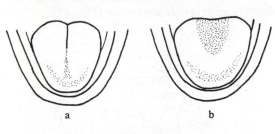

Figure 38

Figure 38   Tongue with (a) and without (b) median furrow

9a Teeth in a single row on edges of jaws .... Mabee's Salamander, *Ambystoma mabeei* Bishop

A small (8–10 cm total length) stocky salamander with many pale flecks (most conspicuous on flanks) on deep brown to black ground color; Coastal plain of N.C. and S.C.

9b Teeth in 3–5 rows on edges of jaws 10

10a Pattern forming narrow pale rings or reticulation .... Flatwoods Salamander, *Ambystoma cingulatum* Cope

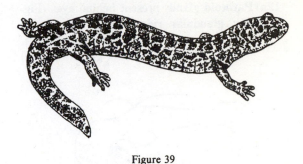

Figure 39

Figure 39   *Ambystoma cingulatum*

Small (9–13 cm total length) black salamanders with pale flecks, frequently forming net pattern or thin rings (Fig. 39); S.C. to northern Fla. and eastern Miss.

10b Pattern absent, or consisting of pale lichen-like flecks; ground color black or very dark brown....... Smallmouth Salamander, *Ambystoma texanum* (Matthes)

Small to moderate sized (11–18 cm total length) with small head and mouth; head narrower than neck. Distribution (Fig. 40).

Figure 40

Figure 40   Distribution of *A. texanum*

11a 10–11 costal folds **12**

11b 12–14 costal folds **13**

12a Parotoid glands present behind eyes (Fig. 41); glandular ridge along top of tail .... Northwestern Salamander, *Ambystoma gracile* (Baird)

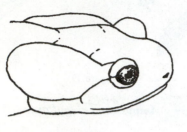

Figure 41

Figure 41   Parotoid glands of *A. gracile*

Moderate sized salamanders (75–115 mm SL); dark blackish brown above, pale brown below; southeastern Alaska south (along coast) to northern Calif.

12b No parotoid glands; no glandular ridge on tail.. .... Mole Salamander, *Ambystoma talpoideum* (Holbrook)

Small (8–12 cm total length), stocky, brown or black salamanders with pale blue-white flecks; tail short; head large. Distribution (Fig. 42).

Figure 42

Figure 42   Distribution of *A. talpoideum*

13a A broad mid-dorsal stripe on body and tail (Fig. 43)... .... Long-toed Salamander, *Ambystoma macrodactylum* Baird

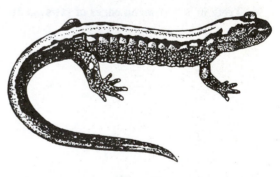

Figure 43

Figure 43   *Ambystoma macrodactylum*

A slender, moderate-sized salamander (100–170 mm total length) with a tan, yellow, or olive dorsal stripe on a black or gray-brown background; southeastern Alaska south to east-central Calif. and east to the Rocky Mountains (western Mont.).

13b   No conspicuous dorsal stripe.............. **14**

14a   Two palmar tubercles; adpressed limbs overlap 1–4 costal folds............................
.... *Ambystoma tigrinum*
Other color variations (see page 24, Fig. 36).

14b   No palmar tubercles; adpressed limbs separated (or overlapping no more than 1 1/2 costal folds).. **15**

15a   Numerous pale spots and flecks on body; area around vent black ....
.... **Blue-spotted Salamander,**
***Ambystoma laterale* Hallowell**

A moderate sized (71–129 mm total length) salamander; bluish-black with white and blue spots. Distribution (Fig. 44).

Figure 44

Figure 44   Distribution of *A. laterale*

Tremblay's Salamander, *Ambystoma tremblayi* Comeau, a triploid ($3n = 42$), apparently all-female species not normally separable from *A.*

laterale. Distribution contained within that of *A. laterale,* best known in southern Mich. and adjacent Ind. and Ohio.

15b   More or less unicolor but sometimes having minute pale flecks on flanks; area around vent gray ... Jefferson Salamander,
***Ambystoma jeffersonianum* (Green)**

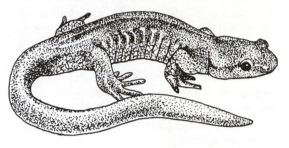

Figure 45

Figure 45   *Ambystoma jeffersonianum*

A moderate-sized (110–210 mm total length) salamander; dark brown to gray above with pale bluish flecks on limbs and flanks (Fig. 45). Distribution (Fig. 46).

Figure 46

Figure 46   Distribution of *A. jeffersonianum*

Silvery Salamander, *Ambystoma platineum* (Cope), a triploid ($3n = 42$), apparently all female species not normally separable from *A. jeffersonianum;* southern Mich., eastern Ind. and western Ohio.

CONGO EEL FAMILY
Amphiumidae

The family Amphiumidae is comprised of one living genus and three species found in the southeastern United States. These elongate salamanders (Fig. 47) superficially resemble eels, snakes, and Sirens but are readily distinguished in having two pairs of tiny legs.

Congo Eels are partially neotenic in that they retain a gill slit as adults. The skull is well ossified and bones appearing late in development (e.g., maxillae, nasals, prefrontals) are well developed.

Fertilization is internal, transfer being accomplished by a spermatophore. Females deposit eggs under debris or in burrows away from standing water and guard the clutch until inundation and hatching.

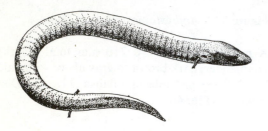

Figure 47

Figure 47 *Amphiuma*

1a Dorsum darker than venter; at least 2 toes on each foot; large salamanders (46–116 cm total length).. 2

1b Dorsum and venter dark; one toe on each foot; small (20–33 cm total length) ....
.... One-toe Amphiuma,
***Amphiuma pholeter* Neill**

A small (normally 22–32 cm total length) dark *Amphiuma* with one toe on each limb. Western Fla. and adjacent Ga.

2a Two toes on each limb; dark dorsal color grades into pale ventral color (Fig. 47)
.... Two-toed Amphiuma,
***Amphiuma means* Garden**

Dorsum darker than venter but grading into pale venter; normally 46–76 cm in total length with 2 toes on each limb. Distribution (Fig. 48).

Figure 48

Figure 48 Distribution of *A. means*

2b Three toes on each limb; dark dorsal color sharply set off from pale venter.................
.... Three-toed Amphiuma,
***Amphiuma tridactylum* Cuvier**

A bicolor (dark above, pale below) salamander normally 46–76 cm in total length with 3 toes on each limb. Found in Miss. drainage from southern Ill. to La. and west to eastern Tex.

GIANT SALAMANDER FAMILY
Cryptobranchidae

A single species of *Cryptobranchus, C. alleganiensis* (Daudin), the Hellbender, is currently recognized as living in the eastern United States. The only other living Giant Salamanders are members of the genus *Andrias,* found in China and Japan. Fossil *Andrias* are known from Europe and North America. Hellbenders are large (to 75 cm in total length), dorsally flattened salamanders with short tails (Fig. 49). Body color is dark gray to brown or black with scattered darker blotches (prominent in the Ozark Hellbender subspecies).

Hellbenders are partially neotenic (one pair of gill slits, eyelids not formed) and aquatic, preferring swiftly flowing rivers and streams. The present distribution is disjunct including the Susquehanna River in Pa. and N.Y., the Ohio River drainage as well as the upper reaches of the Savannah river (Ga., Tenn. and Ala.), and in southern Mo. and northern Ark. (Fig. 50).

Fertilization is external. Eggs, in long rosary-like chains, are deposited beneath rocks or logs and are guarded by adults until hatching occurs.

Figure 49

Figure 49   *Cryptobranchus alleganiensis*

Figure 50

Figure 50   Distribution of *C. alleganiensis*

MUDPUPPY FAMILY
Proteidae

Authorities are divided on whether or not the family Proteidae includes the American salamanders of the genus *Necturus*. The Olm *(Proteus)* is an elongate perennibranch cave-dwelling salamander of Europe and different in many respects from the stocky mudpuppies. All are paedogenetic salamanders lacking derived characteristics and their plesiomorphic similarities are not convincing of a close relationship. Some authorities recognize the family Necturidae for the salamanders of the genus *Necturus*.

Necturus (and the Necturidae) is composed of some five species of salamanders found in the eastern United States and extreme southern Canada. All have large, bushy gills and only four toes on the hind foot. Superficially, they may be confused only with larval and neotenic ambystomatids or dicamptodontids all of which have more slender gill rami and five toes on each hind foot.

Necturus are strongly neotenic. Many skull bones fail to form including the maxillae and nasals.

Fertilization is internal, transfer being accomplished by a spermatophore. Eggs are deposited in long strings beneath logs and rocks and are guarded by the female until hatching.

1a Center of venter not spotted 2

1b Center of venter bearing large or small spots 4

2a Body gray, lacking spots; narrow white band on belly .... Dwarf Waterdog, *Necturus punctatus* (Gibbes)

A small mudpuppy (11–19 cm total length); color gray brown or black, lacking dark spots; throat white; larvae unicolor (no spots or stripes); Coastal plain from Va. to central Ga.

2b Some dark spots on body 3

3a Large spots (larger than a toe) lateral to pale area on venter 5

3b Small spots (no larger than a toe) lateral to pale venter.. .... Alabama Waterdog, *Necturus alabamensis* Viosca

A moderate-sized mudpuppy (15–22 cm total length) variable in coloration; dorsum reddish-brown to slate black with mottling or not; larvae unstriped; western Ga. and northwestern Fla., west to central Miss.

4a Venter bearing many small dark flecks .... Gulf Coast Waterdog, *Necturus beyeri* Viosca

A moderate-sized mudpuppy (16–22 cm total length) dark brown in color with many tan and black spots; larvae spotted; eastern Tex. to central La. and extreme eastern La. to central Miss.

4b Venter bearing several to many large dark spots 5

5a Size small (15–24 cm total length); coastal N.C. .... Neuse River Waterdog, *Necturus lewisi* Brimley

Dorsal and ventral surfaces bearing numerous dark brown or black spots; dorsum rust-brown, venter slate to brown; larvae spotted; Neuse and Tar River systems, N.C.

5b Size larger (adults 20–43 cm total length); not found in coastal N.C. (Fig. 51)............ .... Mudpuppy, *Necturus maculosus* (Rafinesque)

Figure 51

Figure 51   Distribution of *N. maculosus*

Figure 52

Figure 52   *Necturus maculosus*

Gray to rust-brown above with indefinite, round black spots (Fig. 52); venter gray with dark spots; larvae striped.

Newts are primarily a Eurasian family but two genera occur in North America. Fourteen genera and about 45 living species are known in the family whose fossil record is diverse. The American newts are small (5–22 cm total length) animals superficially resembling mole salamanders (Ambystomatidae, Dicamptodontidae) and Lungless salamanders (Plethodontidae) but are easily distinguished in lacking costal folds and grooves. They are likewise distinct in having long parallel rows of teeth (Fig. 14) on the roof of the mouth ("parasphenoid" teeth) and in lacking the nasolabial groove (Fig. 24, characteristic of plethodontids).

Six species of newts are known from the United States. The species of *Taricha* occur in the extreme west (Alaska to southern Calif., an isolate in northern Idaho) whereas the species of *Notophthalmus* occur over the eastern one-half of the United States and in adjacent Canada and northeastern Mexico (Fig. 53).

Newts are unusual among salamanders in that the skin is not so slimy as it is for most amphibians. Eastern newts are distinctive in having a red Eft form (terrestrial juveniles) whereas adults are markedly aquatic. Western newts are terrestrial except during the breeding season when the normally rough skin becomes smooth and the animals migrate to breeding sites. All newts are toxic as food.

Fertilization is internal, transfer being accomplished by a spermatophore. Eggs are deposited in streams (western newts) or ponds (eastern and western) where larvae develop until metamorphosis. Neoteny is uncommon in the family.

Figure 53

Figure 53　Distribution of *Taricha* (western U.S.) and *Notopthalamus* (eastern U.S.)

1a　A pair of nearly parallel ridges on the head extending from snout to occiput (Fig. 54); hind limbs more robust than forelimbs... 2

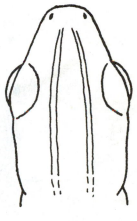

Figure 54

Figure 54　Head of *Notopthalamus*

1b　Top of head lacking cranial ridges; fore and hind limbs equally robust........................ 4

2a Large black spots scattered on head, body, and tail (Fig. 55a) ....
.... **Black-spotted Newt,**
Notophthalmus meridionalis (Cope)

A moderate-sized (71–110 mm total length) newt with a small eye (eye length < eye to nostril distance); olive to dull brown above, deep yellow to orange below; spotted with black; coastal plain of Tex. (Corpus Cristi area) south to northern Veracruz, Mexico.

2b Black spots minute (flecks) 3

3a Eye length equal eye to nostril distance; red stripe extending from eye onto tail (Fig. 55d) .... **Striped Newt,**
Notophthalmus perstriatus (Bishop)

A minute (52–79 mm total length) newt with a large eye (eye length = eye to nostril distance); brown to olive above, yellow below; red dorsolateral stripes usually complete; Northern half of Fla., adjacent Ga.

3b Eye length less than eye to nostril distance; if red markings present, interrupted (Fig. 55b,c) .... **Eastern Newt,**
Notophthalmus viridescens (Rafinesque)

A moderate-sized (65–112 mm total length) newt with a small eye; yellow brown to olive above, yellow below, with many small black spots scattered on dorsum and venter; row of red spots along dorsolateral region or broken pair of red stripes in some subspecies. Distribution (Fig. 56).

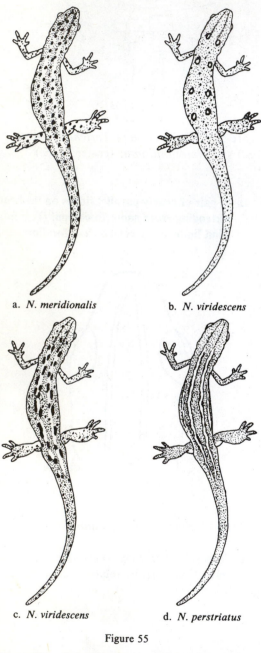

a. *N. meridionalis*   b. *N. viridescens*

c. *N. viridescens*   d. *N. perstriatus*

Figure 55

Figure 55   Color patterns of *Notopthalamus*

Figure 56

Figure 56   Distribution of *Notopthalamus viridescens*

4a Lower eyelid pale (encroached upon by pale ventral coloration).. .... **California Newt,** *Taricha torosa* **(Rathke)**

A large (170 mm total length) newt; dorsum black to dark brown, venter yellow to orange; larvae dark striped; coast ranges of Calif. and western slopes of Sierra Nevada.

4b Lower eyelid dark (encroached upon by dark dorsal coloration)........................... **5**

5a Undersides of limbs dark; iris dark........... .... **Redbelly Newt,** *Taricha rivularis* **(Twitty)**

A large (180 mm total length) newt; brown to black above, bright red below; larvae lacking dark stripes or pale spots; coastal northern Calif.

5b Undersides of limbs pale; iris pale .... **Roughskin Newt,** *Taricha granulosa* **(Skilton)**

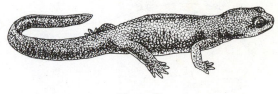

Figure 57

Figure 57   *Taricha granulosa*

A large (175 mm total length) newt (Fig. 57); dark brown to black above, yellow to reddish orange below; larvae bearing pale spots; southeastern Alaska south to west-central Calif.; an isolated population in northern Idaho and adjacent Mont.

SIREN FAMILY
Sirenidae

The family Sirenidae is composed of two living genera and three living species restricted to eastern North America. Superficially, they resemble eels, snakes, and Congo Eels but are readily separated in having external gills and a single pair of limbs (anterior limbs).

In addition to the peculiarity of having no hind limbs, further expression of neoteny is seen in the retention of gills, the absence of eyelids, and the failure of the maxillae to develop (except as rudiments). Premaxillary and maxillary teeth are absent and replaced by keratinous plates.

Fertilization is presumably external although eggs are scattered singly or in small clumps on floating vegetation. Spermatophores are unknown and spermathecae are not present. These are completely aquatic, nocturnal animals.

1a Three toes on each limb; one pair of gill slits; less than 25 cm long (Fig. 58)........... .... **Dwarf Siren,** *Pseudobranchus striatus* **(LeConte)**

Figure 58

Figure 58   *Pseudobranchus striatus*

A minute (12–25 cm total length) siren with tan, gray, black, yellow, buff stripes; southern Ga., the Fla. Peninsula, and adjacent S.C.

1b Four toes on each limb; three pairs of gill slits; usually larger and more robust (18–98 cm) (Fig. 59).. 2

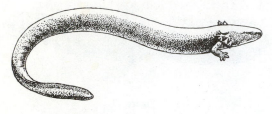

Figure 59

Figure 59   *Siren*

2a Usually 37 costal grooves (36–39) .... Greater Siren, *Siren lacertina* Linnaeus

A large (50–98 cm total length) siren olive to gray in color, usually with faint yellowish dots; venter with pale flecks; Coastal Plain from Md. to western Ala., Fla. peninsula.

2b Usually 33–35 costal grooves (31–38, geographically variable) .... Lesser Siren, *Siren intermedia* LeConte

A medium-sized (18–60 cm total length) siren dark brown to dark black in color. Distribution (Fig. 60).

Figure 60

Figure 60   Distribution of *Siren intermedia*

LUNGLESS SALAMANDER FAMILY
Plethodontidae

Plethodontid salamanders are certainly the most successful group in the order Caudata. The family is composed of 23 genera and more than 200 species. It includes neotenic and troglodytic species as well as fossorial, terrestrial, aquatic, and arboreal species. The vast majority of plethodontids occur in North and South America but the genus *Hydromantes* occurs in the Sierra Nevada of Calif. and in isolated populations in southern Europe.

Plethodontids superficially resemble ambystomatids, dicamptodontids, and salmandrids but differ from each in having nasolabial grooves (Fig. 20). Plethodontids have prominent costal grooves and folds, well developed limbs, and an abundance of teeth on the roof of the mouth (Fig.

14). Plethodontids range in size from minute neotenic or troglodytic forms in Texas (38mm total length) to the Giant Red Hills Salamander of Ala. (256 mm total length). Genera may be conveniently separated on the basis of modifications of the tongue. Several are adetoglossal (the tongue is free all around its perimeter and stands on a pedicel) but most are detoglossal (tongue attached to the floor of the mouth anteriorly) (see Fig. 61).

Life histories are varied. Some, generally considered more primitive taxa, lay eggs in water (streams, bogs, swamps) and have aquatic larvae. Most species of the family lay terrestrial eggs and the larvae metamorphose within the jelly capsule. As is true for most salamanders, fertilization is internal, with sperm transfer being accomplished by a spermatophore. Plethodontids inhabit generally more cool and more moist environments than do ambystomatids. Species densities are highest in the Appalachians, moist forests of extreme western North America, and especially in the neotropics. Living as they do in cooler and more moist environments, their absence of lungs poses no difficulty. The small size of these salamanders means the skin can serve as an extensive organ for gas exchange.

Figure 61

Figure 61   Longitudinal section showing tongue attachment

1a Tail with a basal constriction (Fig. 62).. 2

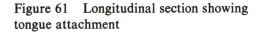

Figure 62

Figure 62   Basal constriction of tail

1b Tail lacking a basal constriction............ 3

2a Four toes on hind feet ....
.... Four-toed Salamander,
Hemidactylium scutatum **(Schlegel)**

A small (51–102 mm total length) salamander with a white belly bearing bold black spots; dorsum rusty-orange with dusky flanks. Distribution (Fig. 63).

Figure 63

Figure 63   Distribution of *Hemidactylium scutatum*

2b Five toes on hind feet................ Ensatina,
Ensatina eschscholtzi **Gray**

Moderate-sized (75–150 mm total length) salamanders exhibiting extreme geographic variation in color; most have yellow to orange limb bases; southwestern British Columbia south to northern Baja Calif., absent from the Great Valley of Calif.

3a Four toes on each hind foot..................... 4

3b Five toes on each hind foot...................13

4a Tail laterally compressed .... **Dwarf Salamander,** *Eurycea quadridigitata* **(Holbrook)**

A small (5–9 cm total length) yellow brown salamander with dark brown dorsolateral stripes; venter yellow-gray. Distribution (Fig. 64).

Figure 64

Figure 64   Distribution of *Eurycea quadridigitata*

4b Tail round in cross-section **5**

5a Venter black with large white spots.......... .... **Oregon Slender Salamander,** *Batrachoseps wrighti* **(Bishop)**

A small salamander (37–56 mm SL) with 16–17 costal grooves, 4.5–7.5 costal folds between addressed limbs; tail 79–118% SL; dorsum dark brown; western slopes of Cascade Mountains in Oreg.

5b Venter lacking white spots **6**

6a Venter white, pale gray, or yellow **7**

6b Venter dark gray to black...................... **8**

7a 5.5–8 costal folds between tips of addressed limbs... ...... **Channel Islands Slender Salamander,** *Batrachoseps pacificus* **(Cope)**

A small salamander (41–63 mm SL) with 18–20 costal grooves; tail 82–104% SL; pale brown above, white to gray below; northern Channel Islands, Santa Barbara County, Calif.

7b 8.5–10.5 costal folds between tips of addressed limbs... .... **Garden Slender Salamander,** *Batrachoseps major* **Camp**

A more slender version of the Pacific Slender Salamander with a longer tail; southwestern Calif. to northern Baja California.

8a 10–12 costal folds between tips of addressed limbs .... **California Slender Salamander,** *Batrachoseps attenuatus* **(Eschscholtz)**

Figure 65

Figure 65   *Batrachoseps attenuatus*

A small salamander (25–50 mm SL) having 19–21 intercostal grooves, tail 98–153% SL (Fig. 65); frequently with a dorsal stripe; venter bearing black network; coastal region from southwestern Oreg. to Baja California; also on Sierra Nevadas of Calif.

8b 2–9.5 costal folds between tips of addressed limbs... **9**

9a 20–21 costal grooves ....
.... **Kern Canyon Slender Salamander,**
Batrachoseps simatus **Brame and Murray**

A small salamander (50–60 mm SL) with 7.5–9
costal folds between tips of adpressed limbs; tail
95–141% SL; venter dark gray-black; flanks
black; dorsum bearing bronze stripe; Kern River
Canyon, Kern County, Calif.

9b 16–19 costal grooves **10**

10a Belly black, undersides of tail flesh colored
.... **Desert Slender Salamander,**
Batrachoseps aridus **Brame**

A small salamander (30–48 mm SL) with 17–18
intercostal grooves; 4–6.5 costal folds between
adpressed limbs; tail 77–90% SL; NW end of
Santa Rosa Mts, Riverside County, Calif.

10b Venter various, not colored as in 10a... **11**

11a 2–5 costal folds between tips of adpressed
limbs ....
.... **Inyo Mountain Salamander,**
Batrachoseps campi
Marlow, Brode, Wake

A small salamander (32–61 mm SL) with 16–18
intercostal grooves and a short tail (tail 62–88%
SL); west slope of Inyo Mountains in northern
Mojave Desert, Inyo County, Calif.

11b 6–9.5 costal folds between tips of adpressed limbs.. **12**

12a Adults small (less than 48 mm SL); 16–19
costal grooves; 7–9.5 costal folds between
tips of adpressed limbs ....
.... **Relictual Slender Salamander,**
Batrachoseps relictus **Brame and Murray**

Adults 33–48 mm SL; dark brown dorsal stripe,
flanks dark black, venter gray-black; western
slopes of Sierra Nevada from Merced River to
Kern Canyon; coastal Monterey and San Luis
Obispo Counties; Santa Cruz Island; San Pedro
Martir Mountains, Baja Calif.

12b Adults larger (51–60 mm SL); 18–19 costal grooves; 6–7 costal folds between tips
of adpressed limbs ....
.... **Tehachapi Slender Salamander,**
Batrachoseps stebbinsi **Brame and Murray**

A salamander with large feet; tail 77–101% SL;
indistinct dorsal stripe; often whitish spots on
ventrolateral surfaces; Piute Range and Tehachapi Mountains, Kern County, Calif.

13a (3) Neotenic Salamanders (gills retained by
sexually mature individuals; often with reduced pigmentation)................................ **14**

13b Adults lack gills...................................... **22**

14a Digits (and limbs) of adpressed limbs overlapping 1–6 costal folds.......................... **15**

14b Digits of adpressed limbs rarely touching,
usually separated by 2–10 costal folds
.... **17**

15a Sides of snout (as seen from above) more or less parallel .... Georgia Blind Salamander, *Haideotriton wallacei* Carr

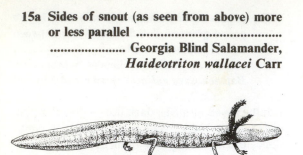

Figure 66

Figure 66   *Haideotriton wallacei*

A small (51–76 mm total length) pinkish-white salamander lacking eye spots and having long slender gills and limbs (Fig. 66); cave systems in southwestern Ga. and adjacent Fla.

15b Sides of snout tapering toward midline (head wedge-shaped) **16**

16a Toes of adpressed limbs overlap 1–4 costal folds .... Comal Blind Salamander, *Eurycea tridentifera* **Mitchell and Reddell**

A small (38–73 mm total length) white or yellow-white salamander with 11–12 costal grooves, long limbs, depressed snout, and concealed eye; Honey Creek Cave, Comal County, Tex.

16b Tips of adpressed limbs overlap 6 costal folds .... Texas Blind Salamander, *Typhlomolge rathbuni* Stejneger

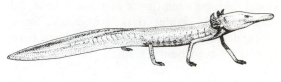

Figure 67

Figure 67   *Typhlomolge rathbuni*

A moderate sized (80–137 mm total length) white salamander with spindly limbs, (Fig. 67), flat snout, minute concealed eye, 12 costal grooves; caves near San Marcos, Tex.

17a Seventeen or more costal folds *and* body width contained 3.5–4 times in interval between limbs .... Tennessee Cave Salamander, *Gyrinophilus palleucus* McCrady

A large, stocky (80–184 mm total length) cave-dwelling salamander; coloration flesh-pink to purplish-brown (subspecies); caves of central and southeastern Tenn. and adjacent Ala. and Ga.

17b 15 or fewer costal folds or if 16 or more, body width contained 4.5 or more times in interval between limbs **18**

18a 19–20 costal grooves .... Oklahoma Salamander, *Eurycea tynerensis* **Moore and Hughes**

A small (44–79 mm total length) gray salamander with a pale venter; at least one row of pale spots along each side of body; northeastern Okla. and adjacent Mo. and Ark.

18b 13–17 costal grooves **19**

19a Eyes minute (Fig. 68a), 13–14 costal grooves ....
.... **Valdina Farms Salamander,**
Eurycea troglodytes **Baker**

Figure 68a

Figure 68a   *Eurycea troglodytes*

A small (51–78 mm total length) pale gray salamander with long slender limbs; northwestern Medina County, Tex.

19b Eyes larger (Fig. 68b), 14–17 costal grooves **20**

20a Body brown or brownish-yellow with rows of pale flecks along flanks **21**

20b Body whitish with faint network of dark pigment ....
.... **Cascade Cavern Salamander,**
Eurycea latitans **Smith and Potter**

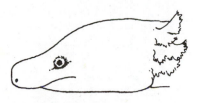

Figure 68b

Figure 68b   *Eurycea latitans*

A small (64–106 mm total length) white salamander with a buried eye and sloping snout; 14–15 costal grooves; caves near Boerne, Kendall County, Tex.

21a Adults minute, 38–51 mm total length .... **San Marcos Salamander,**
Eurycea nana **Bishop**

Minute, brown salamander with 16–17 costal grooves; a row of pale flecks on each side of body; vicinity of San Marcos, Tex.

21b Adults larger, 50–105 mm total length .....
.... **Texas Salamander,**
Eurycea neotenes **Bishop and Wright**

Small, pale brownish-yellow salamander with two rows of pale flecks along each flank; 14–16 costal grooves; Edwards Plateau from vicinity of Austin, Tex., to Val Verde County, Tex.

22a (13) Tongue on a stalk (anterior edge not attached) **23**

22b Anterior edge of tongue attached, lateral and posterior borders free **34**

23a Vomerine teeth form a single series (Fig. 69a) 24

23b Anterior and posterior patches of vomerine teeth (Fig. 69b) 26

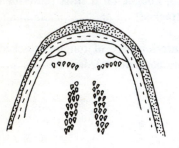

a

b

Figure 69

Figure 69   Vomer teeth arrangements

24a Canthus rostralis absent; 16–17 costal grooves 25

24b Canthus rostralis present, marked by pale line bordered above and below by dark pigment; 17–19 costal grooves .... **Spring Salamander,** *Gyrinophilus porphyriticus* **(Green)**

A large (130–219 mm total length) salamander orange or salmon-colored with diffuse brown flecking or a few small black spots. Distribution (Fig. 70).

Figure 70

Figure 70   Distribution of *Gyrinophilus porphyriticus*

25a Snout relatively longer, equalling 1.5–2 times eye length............ Red Salamander, *Pseudotriton ruber* (Latreille)

A moderate-sized (70–181 mm total length) red, reddish-orange, to purplish salamander with black spots; iris yellow; adults much darker than juveniles. Distribution (Fig. 71).

Figure 71

Figure 71   Distribution of *Pseudotriton ruber*

25b Snout shorter, equalling 1.25–1.5 times length of eye ....
.... **Mud Salamander,**
Pseudotriton montanus Baird

Figure 72

Figure 72   Distribution of *Pseudotriton montanus*

A moderate sized (70–207 mm total length) red salamander with black spots and a brown iris. Distribution (Fig. 72).

26a Toes webbed (Fig. 73a); Cascade and Sierra Nevada Mountains of Calif. 27

Figure 73a

Figure 73a   Toes of *Hydromantes*

26b Toes not webbed (Fig. 73b); eastern U.S. west to Tex. 29

Figure 73b

Figure 73b   Toes of *Eurycea*

27a Dorsum uniform pale to light brown.........
.... **Limestone Salamander,**
Hydromantes brunus Gorman

A moderate-sized (50–64 mm SL) salamander with the toe tips of the adpressed limbs overlapping about 1.5 costal folds; brown above, paler below; Mariposa County, Calif., in Sierra Nevada Mts. at 370–750 meters elevation.

27b Dorsum mottled with dark pigment 28

28a Tips of adpressed limbs overlap 1–2 costal folds ....
.... Shasta Salamander,
***Hydromantes shastae* Gorman and Camp**

A moderate-sized (44–64 mm SL) salamander with slightly less toe webbing than other *Hydromantes;* white patches on venter; northern Calif. south of Shasta Mt. at elevations between 300 and 750 m.

28b Tips of adpressed limbs separated by 1–2 costal folds..
.... Mount Lyell Salamander,
***Hydromantes platycephalus* (Camp)**

A moderate-sized (44–68 mm SL) salamander with olive-gray background flecked with black; venter dusky flecked with white; Sierra Nevada Mts. in east-central Calif. between 1200 and 3350 meters elevation.

29a 19–20 costal grooves ....
.... Many-ribbed Salamander,
***Eurycea multiplicata* (Cope)**

A small (64–97 mm total length) yellowish salamander with darker flanks bearing a row of pale spots; venter bright yellow to gray; northern Ark. and adjacent Mo. and Okla.

29b 14–16 costal grooves 30

30a Tips of adpressed limbs separated by 3–7 costal folds... 31

30b Tips of adpressed limbs separated by 2 or fewer costal folds (sometimes overlapping) 32

31a Tips of adpressed limbs separated by 3–5 costal folds; dorsum pale with dark dorsolateral stripes or rows of spots ....
.... Two-lined Salamander,
***Eurycea bislineata* (Green)**

A small (64–121 mm total length) yellow salamander with bold brown to black markings on flanks; venter yellow. Distribution (Fig. 74).

Figure 74

Figure 74   Distribution of *Eurycea bislineata*

31b Tips of adpressed limbs separated by 5–7 costal folds; dark brown without dorsolateral stripes (Tex.—metamorphosed individuals) .... Texas Salamander
Eurycea neotenes
see 21b

32a Tail length 47–52% total length; brown above with dark flecks forming dorsolateral lines .... Junaluska's Salamander,
Eurycea junaluska
Sever, Dundee and Sullivan

A small (34–44 mm SL) brown salamander with lateral bands of flecking; 0–1 costal folds between adpressed limbs; Cheoah River valley, Graham County, N.C.

32b Tail length 51–65% total length; variously pigmented but not brown 33

33a Pattern of small black spots, not more intense on flanks than dorsum .... **Cave Salamander,** *Eurycea lucifuga* **(Rafinesque)**

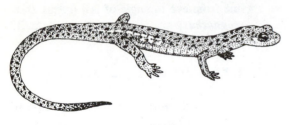

Figure 75

Figure 75   *Eurycea lucifuga*

A moderate-sized (100–181 mm total length) yellow to orange salamander spotted with black (Fig. 75); venter yellow, normally unspotted. Distribution (Fig. 76).

Figure 76

Figure 76   Distribution of *Eurycea lucifuga*

33b Pattern of brown spots, usually forming more intense pattern on flanks than on dorsum .... **Longtailed Salamander,** *Eurycea longicauda* **(Green)**

A moderate-sized (90–200 mm total length) dull yellow to orange salamander with the flanks darker than the dorsum; subspecies distinctive in coloration; long tail. Distribution (Fig. 77).

Figure 77

Figure 77   Distribution of *Eurycea longicauda*

34a (22) Elongate salamander, more than 12 costal folds between tips of adpressed limbs .... **Red Hills Salamander,** *Phaeognathus hubrichti* **Highton**

A large (100–256 mm total length) dark brown salamander, 20–22 costal grooves, with short legs; southern Ala., Red Hills formation.

34b Adpressed limbs overlap or separated by many fewer (usually no more than 4–6) costal folds than 12 .... **35**

35a Choanae slit-like (often not visible) (Fig. 78a) ....
.... Shovelnose Salamander, *Leurognathus marmoratus* Moore

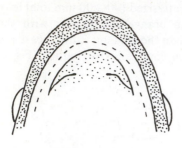

Figure 78a

Figure 78a   Slit-like choanae in *Leurognathus*

A moderate-sized (90–146 mm total length) stocky salamander with more robust hind than forelimbs; 13–14 costal grooves; dorsum dark (gray, brown, or black) with two rows of indistinct pale spots; aquatic; southwestern Va. to northeastern Ga. at elevations between 300 and 1700 m.

35b Choanae round, easily visible on roof of mouth (Fig. 78b) 36

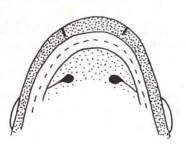

Figure 78b

Figure 78b   Round choanae

36a Fore and hind limbs about equal in size; no white stripe from eye to angle of jaws ......
.... 45

36b Hind limbs larger than forelimbs; a white line from eye to angle of jaw (genus *Desmognathus*) 37

37a Tail rounded above 38

37b Tail keeled above (sometimes only toward posterior end)......................... 40

38a Minute forms, adults less than 60 mm total length 39

38b Larger salamanders, adults 70–110 mm total length ....
.... Mountain Dusky Salamander, *Desmognathus ochrophaeus* Cope

Coloration highly variable, most have zig-zag dorsal stripe but in some this exists as a row of blotches or as a stripe containing many darker chevrons; brown to orange above, darker on flanks; upland areas from northeastern N.Y. to northern Ga. and adjacent Ala.

39a Top of head rugose; dorsal stripe bearing many chevrons ....
.... Pigmy Salamander, *Desmognathus wrighti* King

Minute (38–51 mm total length) salamanders with tan to reddish brown dorsal stripe; mental gland large, U-shaped; upland areas along border between Tenn. and N.C.

39b Top of head smooth; dorsal stripe various .... Seepage Salamander, *Desmognathus aeneus* **Brown and Bishop**

Minute (44–57 mm total length), slender salamander with small, kidney-shaped mental gland; southwestern N.C. to east-central Ala.; also in west-central Ala.

40a Venter mottled light and dark **41**

40b Venter without mottling (*either* light *or* dark) **43**

41a Tips of toes dark .... **Black Mountain Salamander,** *Desmognathus welteri* **Barbour**

A moderate-sized (80–170 mm total length) chunky salamander; above, small dark brown spots on brown ground color; belly whitish, stippled with gray or brown pigment; eastern Ky.

41b Tips of toes pale **42**

42a Pale spots along lower flanks .... **Southern Dusky Salamander,** *Desmognathus auriculatus* **(Holbrook)**

Moderate-sized (80–163 mm total length) salamander; dark brown or black above, venter same but bearing distinct white spots; Coastal Plain from southeastern Va. to central Fla. and west to eastern Tex.

42b Lower flanks flecked or not, lacking distinct pale spots .... **Dusky Salamander,** *Desmognathus fuscus* **(Rafinesque)**

A moderate-sized (60–141 mm total length) salamander highly variable in coloration. Distribution (Fig. 79).

Figure 79

Figure 79 Distribution of *Desmognathus fuscus*

43a Venter black, 2 rows of light spots along flanks .... **Blackbelly Salamander,** *Desmognathus quadramaculatus* **(Holbrook)**

Figure 80

Figure 80 *Desmognathus quadramaculatus*

A large (100–210 mm total length) robust salamander (Fig. 80) with a short tail (less than ½ total length); south-central W.Va. to northern Ga. in upland areas.

43b Venter pale **44**

44a Heavy dark markings on dorsum ....
.... **Seal Salamander,**
Desmognathus monticola **Dunn**

A moderate-sized (80–150 mm total length) robust salamander; gray to brown above with reticulation or network of black or dark brown
markings; southwestern Pa. to eastern Tenn. and
adjacent N.C. and Ga.; isolated populations in
central Ala. and adjacent Ga.

44b Dorsum brown to gray with 2 rows of faint
pale spots along lateral edges of back
.... **Ouachita Dusky Salamander,**
Desmognathus brimleyorum **Stejneger**

A moderate-sized (80–180 mm total length) robust salamander; belly pinkish-white to yellowish; Ouachita Mountains of western Ark. and
eastern Okla.

45a (36) Light and dark stripes on lower flanks
....
.... **Many-lined Salamander,**
Stereochilus marginatus **(Hallowell)**

A small (64–114 mm total length) brown salamander with streaks along lower flanks; limbs
short, 7–9 costal folds between adpressed limbs,
16–18 costal folds; Coastal Plain from southern
Va. to eastern Ga.

45b Not so colored **46**

46a Eyes small, partially concealed (covered by
fused eyelids)...
.... **Grotto Salamander,**
Typhlotriton spelaeus **Stejneger**

A small (80–135 mm total length) whitish salamander with concealed eyes (evident as dark
spots); 16–19 costal grooves, about 4 between tips
of adpressed limbs; caves of southern Mo. and
adjacent Ark. and Okla.

46b Eyes normal; eyelids not fused **47**

47a Maxilla lacking teeth on posterior end
(below eye); most have truncate toe tips
(slightly expanded); premaxillary teeth en-
larged (genus *Aneides*)...........................**48**

47b Maxilla toothed posterior to eye (genus
***Plethodon*)** **52**

48a Dorsum black with yellow to green lichen-
like markings ....
.... **Green Salamander,**
Aneides aeneus **(Cope)**

A moderate-sized (80–140 mm total length)
slender salamander; 14–15 costal folds; toe tips
of adpressed limbs overlap 1–3 costal folds; Appalachians from southwestern Pa. to northern
Ala.

48b Not so colored, western North America (Fig. 81)................................ **49**

Figure 81

Figure 81   Distribution of *Aneides* species

49a Tips of adpressed limbs separated by 2–5 costal folds................................ **50**

49b Tips of adpressed limbs separated by no more than 1.5 costal folds (may overlap) **51**

50a Black above and below .... **Black Salamander,** *Aneides flavipunctatus* (Strauch)

A moderate-sized to large salamander (60–93 mm SL) with 14–16 costal grooves; adpressed limbs separated by 3–5 costal folds; white and/or brassy iridophores scattered on dorsum; northern Calif.

50b Brown above, pale below .... **Sacramento Mountain Salamander,** *Aneides hardyi* (Taylor)

Adults moderate-sized (40–58 mm SL) with 14–15 costal grooves; adpressed limbs separated by 2–4 costal folds; dorsum mottled with bronze; 2590–3360 m in the Capitan Mts., Sierra Blanca, and Sacramento Mts., in Lincoln and Otero Counties, N. Mex.

51a 16–17 costal grooves; not spotted with yellow .... **Clouded Salamander,** *Aneides ferreus* Cope

Adults moderate-sized (45–75 mm SL); adpressed limbs separated by no more than 1.5 costal folds (often overlapping); dark brown mottled with brassy to gray; venter whitish; Vancouver Island, British Columbia; coastal areas from Columbia River in Oreg. to northern Calif.

51b 15–16 costal grooves; spotted with yellow (Fig. 82).. .... **Arboreal Salamander,** *Aneides lugubris* (Hallowell)

Figure 82

Figure 82   *Aneides lugubris*

A moderately large (65–100 mm SL) salamander; tips of adpressed limbs overlap one or more costal folds; brown above with many yellow spots; venter white to gray; coastal Calif. to northern Baja California; western slopes of Sierra Nevada Range.

52a (47) Western species (found west of the Great Plains) (Fig. 83) **53**

52b Eastern species (found as far west as central Tex.) (Fig. 83) **60**

Figure 83

Figure 83   Distribution of *Plethodon* species

53a Normally 14 costal grooves .... **Van Dyke's Salamander, *Plethodon vandykei* Van Denburg**

A moderate-sized salamander (45–65 mm SL) with an uneven-edged dorsal stripe bordered by black or dark brown; throat pale yellow; western Wash., northern Idaho, and adjacent Mont.

53b 15–19 costal grooves **54**

54a Costal grooves 15–16 .... **55**

54b Costal grooves 17–19 .... **58**

55a No dorsal stripe; 2.5 costal folds between tips of adpressed limbs .... **Mary's Peak Salamander, *Plethodon gordoni* Brodie**

Moderate-sized (50–72 mm SL); dark gray to brown with white flecks; Benton, Lane, and Lincoln Counties, Oreg.

55b Dorsal stripe present, *or if not,* more than 2.5 costal folds between tips of adpressed limbs **56**

56a Dorsal stripe with uneven edges (Fig. 84a) **57**

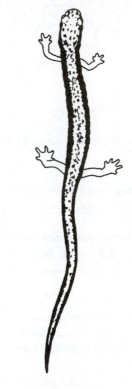

Figure 84a

Figure 84a   *Plethodon dunni*

**56b Dorsal stripe with even edges (Fig. 84b) ...
.... Western Redback Salamander,
Plethodon vehiculum (Cooper)**

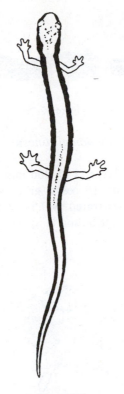

Figure 84b

Figure 84b *Plethodon vehiculum*

Moderate-sized (45–62 mm SL); 3–5 costal folds between tips of adpressed limbs; dorsal stripe extends to tip of tail; Vancouver Island to southwestern Oreg.

**57a One phalanx in fifth toe of hind foot; no
mental gland in males.................................
.... Larch Mountain Salamander,
Plethodon larselli Burns**

Moderate-sized (40–57 mm SL) salamanders; 15 costal grooves; 0–3 costal folds between tips of adpressed limbs; dorsal stripe usually reaching tip of tail; venter red-orange; lower Columbia River gorge between Hood River and Troutdale, Oreg.

**57b Two phalanges in fifth toe; mental gland in
males (Fig. 85) ....
.... Dunn's Salamander,
Plethodon dunni Bishop**

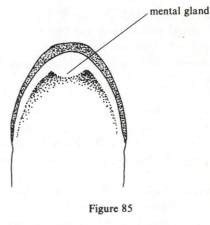

mental gland

Figure 85

Figure 85 *Plethodon dunni*

Moderate-sized (50–75 mm SL) salamanders with 15 costal grooves; greenish-yellow dorsal stripe (not reaching tail tip); venter slate-colored; coastal Oreg.

58a 19 costal grooves; fifth toe shortened........ .... Jemez Mountains Salamander, *Plethodon neomexicanus* Stebbins and Reimer

A slender, moderate-sized (50–75 mm SL) salamander with short legs, 7.5–8.5 costal folds between tips of adpressed limbs; brown with brassy stippling; Jemez Mts., N. Mex.

58b 17–18 costal grooves; fifth toe not reduced in size... **59**

59a A brown salamander flecked with white.... .... Siskiyou Mountains Salamander, *Plethodon stormi* Highton and Brame

A moderate-sized (55–76 mm SL) salamander with 17 costal grooves and 4–5.5 costal folds between tips of adpressed limbs; venter lavender, throat cream; Siskiyou County, Calif., and Jackson County, Oreg.

59b Dorsum brown or rust, not flecked with white .... Del Norte Salamander, *Plethodon elongatus* Van Denburgh

A moderate-sized (55–73 mm SL) salamander with 18 costal grooves and usually 6.5–7.5 costal folds between tips of adpressed limbs; venter dark gray, throat paler; southwestern Oreg. and northwestern Calif.

60a (52) 16 costal grooves **61**

60b 17–22 costal grooves **67**

61a Dorsum black with white spots (Fig. 86); venter slate (throat pale in Texas populations) .... Slimy Salamander, *Plethodon glutinosus* (Green)

Figure 86

Figure 86 *Plethodon glutinosus*

A moderately large (120–206 mm total length) salamander with 16 costal grooves; tips of adpressed limbs separated by 0.5 to 3.5 costal folds; throat dark. Distribution (Fig. 87).

Figure 87

Figure 87 Distribution of *P. glutinosus*

61b Dorsum various, if black with pale spots, venter not uniformly slate-colored **62**

62a 3–4 costal folds between tips of adpressed limbs; venter black, sometimes with white spots .... Weller's Salamander, *Plethodon welleri* Walker

A small (64–79 mm total length) salamander with 16 costal grooves; gold to silvery blotches above; extreme northeastern Tenn. and adjacent N.C. and Va.

62b 3 or fewer costal folds between adpressed limbs; venter various but not black....... 63

63a Limbs long, tips of adpressed limbs overlapping to separated by no more than one costal fold 64

63b Limbs shorter, separation of 1–3 costal folds 65

64a Light lateral stripe edged above by black .... Crevice Salamander, *Plethodon longicrus* **Adler and Dennis**

A large (130–221 mm total length) salamander with flecks or scattered blotches of chestnut pigment dorsally; 16 costal grooves; Rutherford County, N.C.

64b Light lateral stripe edged above by chestnut or red .... Yonahlossee Salamander, *Plethodon yonahlosee* **Dunn**

A large (110–190 mm total length) salamander with a broad red or chestnut stripe down the back; 16 costal grooves; southern Blue Ridge Mts. (southwest Va. to Tenn.).

65a Numerous white flecks in dorsal pattern; 0–1 costal folds between adpressed limbs 66

65b White flecks rarely present; pattern extremely variable (geographically) .... Appalachian Woodland Salamander, *Plethodon jordani* **Blatchley**

A moderately large (90–184 mm total length) salamander with 16 costal grooves; 1–3 costal folds between adpressed limbs; most individuals are black without markings but others have white

flecks on the dorsum and red spots on the legs; one variety has red cheeks and another has red legs; highlands from southwestern Va. to northeastern Ga.

66a Chest dark; north of Ouachita River and west of a line south from Mena .... Rich Mountain Salamander, *Plethodon ouachitae* **Dunn and Heinze**

A moderate-sized salamander (100–159 mm total length) with 16 costal grooves; throat whitish, chest dark; pattern variable; Ouachita Mts., Ark. and Okla.

66b Chest pale, flecked with black and white; south of Ouachita River and east of a line south from Mena .... Caddo Mountain Salamander, *Plethodon caddoensis* **Pope and Pope**

A small to moderate-sized salamander (90–111 mm total length) with 16 costal folds; black and white above; throat light, chest peppered with black and white; Caddo Mts. of western Ark.

67a (60) Belly unicolor (gray to black)........ 68

67b Belly mottled with dark and light pigment 71

68a 19–21 costal grooves .... Ravine Salamander, *Plethodon richmondi* **Netting and Mittleman**

A moderate-sized (80–143 mm total length) slender salamander with 5–10 costal folds between adpressed limbs; brown to black above with minute bronze flecks; throat dark with white speckling; western Pa. to southeastern Ind. and south to extreme northeastern Tenn.

68b 17–18 costal grooves 69

69a Flanks black with gold flecks; 5–6 costal folds between tips of adpressed limbs .... **Cheat Mountain Salamander,** *Plethodon nettingi* **Green**

A small to moderate-sized (80–122 mm total length) salamander with 18 costal grooves; dorsum black with brassy flecks; venter gray to black; eastern W. Va. and adjacent Va.

69b Flanks bearing white spots; 2–4 costal folds between tips of adpressed limbs 70

70a White spots on dorsum; 17–18 costal grooves .... **White-spotted Salamander,** *Plethodon punctatus* **Highton**

A moderate-sized salamander (100–157 mm total length) lacking brassy flecking and red spots; Va.–W. Va. line.

70b 17 costal grooves .. **Wehrle's Salamander,** *Plethodon wehrlei* **Fowler and Dunn**

A moderate-sized salamander (100–160 mm total length); dark brown with row of whitish spots along each side of body; often red spots in mid-dorsal region; southwestern N.Y. to Va. and adjacent N.C.

71a Throat mostly white, contrasting sharply with darker venter.................................... .... **Valley and Ridge Salamander,** *Plethodon hoffmani* **Highton**

A small to moderate-sized salamander (80–137 mm total length) with 21 costal grooves; dorsum black with pale flecks; along frontier between Va. and W. Va. northeasterly to Susquehanna River Valley of Pa.

71b Throat mottled like venter but paler than venter 72

72a Costal grooves 18–20, usually 19 (but geographically variable); dorsal stripe, if present, straight-edged and narrowing slightly at base of tail (Fig. 88) .... **Redback Salamander,** *Plethodon cinereus* **(Green)**

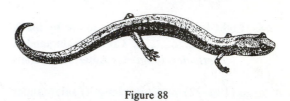

Figure 88

Figure 88   *Plethodon cinereus*

A small to moderate-sized salamander (57–127 mm total length) with 6–9 costal folds between adpressed limbs; either "red-backed" color phase or the "lead-back" phase; venter mottled black and white. Distribution (Fig. 89).

Figure 89

Figure 89   Distribution of *P. cinereus*

72b Costal grooves 18; dorsal stripe with wavy or zig-zag edges, or very narrow and not narrowing at base of tail ....
.... **Zigzag Salamander,**
Plethodon dorsalis Cope

A small to moderate-sized salamander (60–111 mm total length) with 6–7 costal folds between adpressed limbs; dorsal stripe red to yellowish; venter mottled black and orange to reddish; cen-

tral Ind. to western Ga. and northern Ala. Isolated populations peripheral to main range.

Southern Redback Salamander, *Plethodon serratus* Grobman; this small salamander will either key out as *P. cinereus* or *P. dorsalis*. The dorsal stripe is of uniform width and has serrated edges. Distributed in the Ozarks of Mo. and in isolated populations in Ark., La. and Okla.

FROGS AND TOADS
Salientia (Anura)

Frogs and toads are monotonous in many features. All have the elongate ilia and hind limbs. Proportions among various sorts of frogs range from stocky, short-legged forms (often termed toads) to more gracile, long-legged forms (frogs and tree frogs). Much of their classification is based on osteological features which are not often evident or readily accessible when attempting to identify specimens. For typical features see Fig. 90.

The head and trunk are not separated from one another in many frogs but in most the body narrows briefly immediately behind the head. Some species of toads bear bony ridges on the head (cranial crests). Other ridges and folds on the body useful in species discrimination include the transverse head fold and dorsolateral folds. Toads have prominent glands located behind the eye (parotoid glands) and some have enlarged glands on the shank. A variety of skin textures is encountered among frogs ranging from perfectly smooth to coarsely tuberculate or warty. The skin of the belly may be smooth or granular. In most of our native frogs, a tympanum is evident posterior to the eye but in some species the structure is hidden beneath the skin and in a few completely absent. The pupil is horizontal or vertical.

Of considerable value are features evident on the ventral surfaces of the hands and feet.

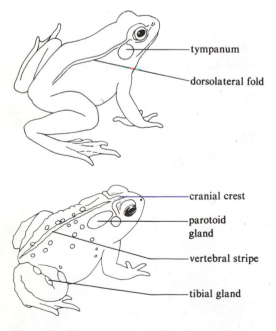

Figure 90   Typical structures of frogs and toads

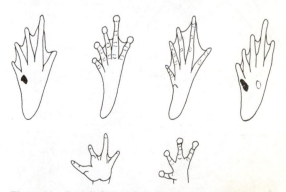

Figure 91   Feet (left to right): *Scaphiopus, Hyla, Rana, Bufo*, Hands (left to right): *Rana, Hyla*

Webbing of varying degrees is developed between the digits of many frogs (Fig. 91). The fingers and toes may be simple pointed structures or may bear distal expansions (pads). The pads usually bear adhesive discs on their ventral surfaces. In tree frogs (Hylidae) an accessory phalanx, the intercalated cartilage, occurs between the terminal phalange and more proximal phalanges (Fig. 92). It is most easily detected in lat-

Figure 92   Intercalary cartilage of Hylidae

eral profile. The Clawed Frogs bear epidermal claws on the inner three toes of the hind foot.

In addition to digital pads and webbing, the feet of frogs differ in the presence or absence of the outer metatarsal tubercle. The inner metatarsal tubercle is larger than the outer and elongate but may be heavily keratinized and spadelike. Proximal to the inner metatarsal tubercle, the tarsus may bear a fold (inner tarsal fold).

Inside the mouth (Fig. 93), the palate may bear vomerine odontophores between and/or

posterior to the choanae. The maxillary arch normally bears teeth on the premaxillae and maxillae. These are small but can be detected by drawing a pin along the inner edge of the lip. No teeth are borne on the lower jaw. The tongue is normally attached at the front and free along its posterior edge. The posterior edge is either deeply notched (with cornua) or round. Vocal slits (if present) lie posterolateral to the tongue.

The pectoral girdle is either arciferal or firmisternal. In arciferal frogs, the two halves of the pectoral girdle overlap on the midline whereas in firmisternal frogs the two halves are fused on the midline (Fig. 94).

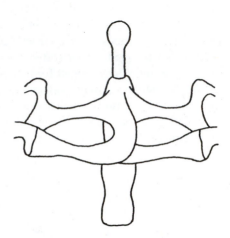

Figure 94a   Arciferal pectoral girdle

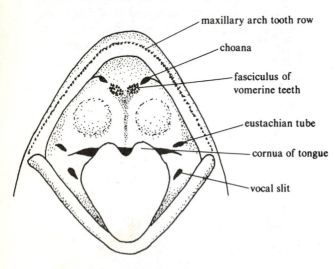

maxillary arch tooth row

choana

fasciculus of vomerine teeth

eustachian tube

cornua of tongue

vocal slit

Figure 93   Structures of frogs mouth

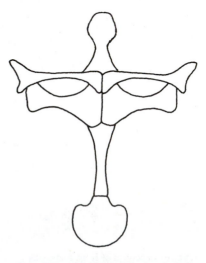

Figure 94b   Firmisternal pectoral girdle

KEY TO FAMILIES OF FROGS AND TOADS

1a Toes fully webbed; inner three toes on hind foot bearing black claws: tongue absent (Fig. 95)..................... (p. 58) TONGUELESS FROG FAMILY, Pipidae

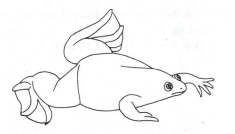

Figure 95   Clawed frog

1b Tongue present; toes lacking claws; webbing variable 2

2a Only four toes evident on hindfoot (first forming another inner metatarsal tubercle); tongue free at anterior end (Fig. 96).. .... (p. 57) BURROWING TOAD FAMILY, Rhinophrynidae

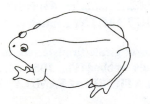

Figure 96   Burrowing toad

2b Five toes evident on hind foot; tongue attached at anterior end 3

3a Pupil a vertical slit or ellipse (Fig. 97a); outer metatarsal tubercle absent............. 4

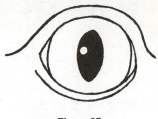

Figure 97a

Figure 97a   Vertical pupil of eye

3b Pupil a horizontal ellipse (Fig. 97b); outer metatarsal tubercle usually present........ 5

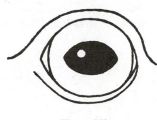

Figure 97b

Figure 97b   Horizontal pupil of eye

4a Large inner metatarsal tubercle cornified (black or brown) (Fig. 98) .... (p. 60) ARCHAIC TOAD FAMILY, Pelobatidae

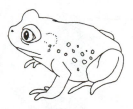

Figure 98   Archaic toad

4b Small inner metatarsal tubercle not cornified (Fig. 99)... .... (p. 57) TAILED FROG FAMILY, Ascaphidae

Figure 99   Tailed frog

Figure 102   Narrow-mouthed toad

5a Upper jaw lacking teeth 6

5b Upper jaw toothed 7

7a Digits bearing intercalary cartilages (Fig. 92) (Fig. 103) .... (p. 73) **TREE FROG FAMILY, Hylidae**

6a Parotoid glands present posterior to eye (Fig. 100) .... **(p. 66) TOAD FAMILY, Bufonidae**

Figure 100   Toad

Figure 103   Tree frog

7b Digits lacking intercalary cartilages 8

6b Parotoid glands absent; an occipital fold across head (Fig. 101) (Fig. 102) .... **(p. 58) NARROW-MOUTHED TOAD FAMILY, Microhylidae**

8a Pectoral girdle arciferal; toes not webbed (Fig. 104) .... **(p. 63) NEOTROPICAL FROG FAMILY, Leptodactylidae**

8b Pectoral girdle firmisternal; toes partially to fully webbed (Fig. 105) .... **(p. 82) AQUATIC FROG FAMILY, Ranidae**

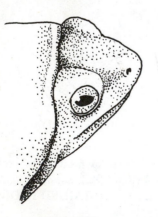

Figure 101

Figure 101   Occipital fold on Microhylid

Figure 104   Neotropical frog

Figure 105   Aquatic frog

TAILED FROG FAMILY
Ascaphidae

The limits of this family are uncertain. Only the Bell Toad, *Ascaphus truei,* is included here but some authorities also include three species of New Zealand frogs (genus *Leiopelma*). Two fossil (Jurassic) frogs (*Notobatrachus* and *Vieraella*) from Argentina are also members of the Ascaphidae.

The Bell Toad (Fig. 106) is a small, 25–50 mm SVL (=snout-vent length, distance from tip of snout to anal opening), warty frog with fully-webbed feet distributed west of the Cascade Mountains from extreme southern British Columbia southward into northern Calif. and in the Rocky Mountains in northern Idaho, adjacent Mont., southeast Wash., and northeast Oreg. These frogs inhabit cool, clear streams in forests. Unlike most frogs, fertilization is internal. The "tail" is a cloacal extension found in males and serves as an intromittant organ. The tadpoles are adapted to stream life. The large sucker-mouth is ventral and provided with many rows of den-

ticles. The spiracle lies on the throat. Eggs are deposited beneath stones in strings. Development is slow, 2–3 years being required for the larval period. Amplexus is inguinal.

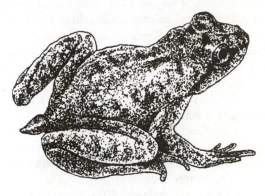

Figure 106

Figure 106   *Ascaphus truei*

BURROWING TOAD FAMILY
Rhinophrynidae

Rhinophrynus dorsalis (Mexican Burrowing Frog) is a peculiar toad (Fig. 107). Although juveniles may be nearly entirely black, orange marks are normally seen on the digits and dorsum. The small head and globular body make these frogs seem quite similar to narrow-mouthed toads but the burrowing toad is readily distinguished in having only four toes on its hind feet (the innermost toe is modified into another metatarsal tubercle). Adults are moderate-sized, 50–90 mm SVL, and feed on ants and termites.

Rhinophrynids have larvae very similar to those of pipid frogs. The larvae lack denticles and beaks, have a pair of spiracles, and have numerous short barbels (4 pair of lateral barbels, one pair of dorsal barbels, and one ventral barbel) around the mouth. Larvae school, swim head down, and use the tip of the tail for propulsion. Amplexus is inguinal and fertilization is external. Reproduction occurs in ponds.

The only living rhinophrynid is distributed in lowlands from southern Tex. to western Costa Rica. Early Tertiary fossils are known from Canada and Wyo.

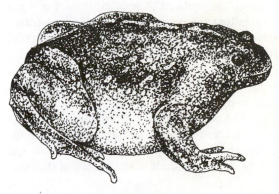

Figure 107

Figure 107   *Rhinophrynus dorsalis*

TONGUELESS FROG FAMILY
Pipidae

Pipids are native to Africa and South America. The neotropical representatives are included in *Pipa* whereas the African species are partitioned into 3 genera. The African frogs are called clawed frogs because the three innermost toes bear epidermal claws. *Xenopus*, (Fig. 108) the largest of the African frogs, is commonly sold in pet stores. In the Paleocene, *Xenopus* occurred in Brazil. Presumably as a result of release of pets, *Xenopus laevis* is established in stream systems in southern Calif.

Adults reach 100 mm SVL. This is a totally aquatic frog which moves awkwardly on land. The tongue is absent. Lateral line organs are well-developed in adults. The toes are fully webbed. These frogs are dark brown above with drab olive spots. The venter is dirty-cream.

Larvae are pelagic filter-feeders lacking horny beaks and denticles. Larvae have a pair of ventrolateral spiracles. The larvae resemble small catfishes in having a long filamentous barbel at each corner of the mouth. Larvae swim in schools with the head angled downward in the water. The tip of the tail undulates providing slow propulsion. During rains adults migrate from pond to pond. Amplexus is inguinal and fertilization is external. Eggs are scattered about on vegetation or on the pond bottom.

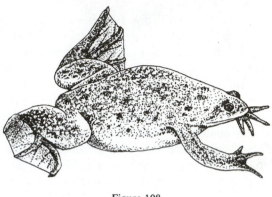

Figure 108

Figure 108   *Xenopus laevis*

NARROW-MOUTHED TOAD FAMILY
Microhylidae

This large (225 species, 58 genera, 8 subfamilies) pantropical family of toads extends into temperate regions in North America. All American genera are closely related. A peculiar feature of these frogs is that the tympanum is hidden by a transverse fold of skin across the back of the head (Fig. 101). Microhylids are toothless (like true toads) and feed on ants and termites. The small head and snout are pointed, contrasting with the plump body and short limbs.

The pectoral girdles of microhylids are firmisternal and there is a marked tendency towards reduction of girdle elements within the family. About one-half of the genera of the family have a peculiar articulation between the eighth and ninth vertebrae found elsewhere among true frogs (ranids). Microhylids and ranids also share a peculiar pattern of muscle attachments in the thigh.

Microhylids engage in axillary amplexus (except for certain peculiar African forms) and fertilization is external. Most microhylids have free swimming tadpoles but others abbreviate development. Larvae are pelagic with lateral eyes; beaks and denticles are absent from the minute terminal mouth. Larvae have a single ventral spiracle.

1a Two metatarsal tubercles ..... Sheep Frog,
Hypopachus variolosus **(Cope)**

Figure 109

Figure 109 *Hypopachus variolosus*

A small (25–44 mm SVL) plump frog (Fig. 109); dorsum brown (reddish brown laterally) with yellow vertebral streak; venter mottled with brown and bearing thin yellow lines. Extreme southern Tex. southeasterly to Costa Rica.

1b One metatarsal tubercle 2

2a Venter tan to olive green, not mottled with browns..
.... Great Plains Narrowmouth Toad,
Gastrophryne olivacea **(Hallowell)**

A small (22–41 mm SVL) plump frog; some juveniles have dark markings (leaf-shape) on dorsum; adults are tan to olive green with occasional black flecks. Distribution (Fig. 110).

Figure 110

Figure 110 Distribution of *Gastrophryne olivacea*

2b Venter heavily mottled with browns..........
.... Eastern Narrowmouth Toad,
Gastrophryne carolinensis **(Holbrook)**

A small (22–38 mm SVL) plump frog with a brown dorsum (flecked with black) and yellow-tan dorsolateral bands. Distribution (Fig. 111).

Figure 111

Figure 111 Distribution of *G. carolinensis*

ARCHAIC TOAD FAMILY
Pelobatidae

The eight living genera and 49 living species of Archaic Toads are currently divided into two subfamilies. Megophryines are found in the Oriental Region (Himalayas to the Philippines) whereas pelobatines occur in Europe, western Asia, northern Africa, and North America. Unlike most frogs and toads, the Archaic Toads have a rich fossil history.

Archaic Toads have arciferal pectoral girdles, vertical pupils, and lack an outer metatarsal tubercle. The skin is smooth and handling results in production of a copious and pungent secretion (which may elicit an allergic reaction). Maxillary and premaxillary teeth are present. In pelobatines, the sacrum is fused to the coccyx.

Amplexus is inguinal and fertilization is external. Ovoposition occurs in ponds and other temporary bodies of water. Larvae have horny beaks and numerous rows of labial denticles. The spiracle is sinistral.

The American Archaic Toads are here considered members of the genus *Scaphiopus*. Some authorities recognize two genera. In that event, *Scaphiopus* is restricted to forms having more massive skulls and sickle-shaped metatarsal tubercles (*S. couchi* and *S. holbrooki*) and *Spea* is used for the other species (*S. bombifrons, S. hammondi, S. intermontanus,* and *S. multiplicatus*).

1a Metatarsal tubercle sickle-shaped (Fig. 112a).. 2

1b Metatarsal tubercle wedge-shaped (Fig. 112b).. 3

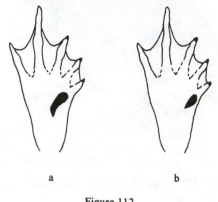

a   b

Figure 112

Figure 112   Shape of metatarsal tubercle

2a Tympanum indistinct .... **Couch's Spadefoot,** *Scaphiopus couchi* **Baird**

Figure 113

Figure 113   *Scaphiopus couchi*

Moderate-sized (57–90 mm SVL) toad yellowish above marbled with black, brown or green (Fig. 113); no boss between eyes; interorbital space as wide as upper eyelid; central Tex. and southern Okla. west to southeastern Calif. and south to central Mexico.

2b Tympanum distinct.... Eastern Spadefoot, *Scaphiopus holbrooki* **(Harlan)**

Moderate-sized (44–83 mm SVL) toad brown above with yellow paravertebral stripes (lyre-shaped pattern). Distribution (Fig. 114).

Figure 114

Figure 114   Distribution of *S. holbrooki*

3a A boss (bony or glandular) between eyes (Fig. 115)... **4**

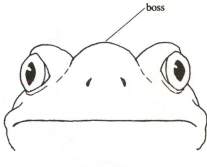

boss

Figure 115

Figure 115   Boss on head of *S. bombifrons*

3b No boss between eyes ....
.... **Western Spadefoot,**
Scaphiopus hammondi **Baird**

A moderate-sized (38–64 mm SVL) toad with reddish tubercles on a gray or greenish to dark brown dorsum; eyelid wider than interorbital space. Distribution (Fig. 116).

Figure 116

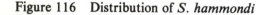
Figure 116   Distribution of *S. hammondi*

4a Boss bony....................... **Plains Spadefoot,**
Scaphiopus bombifrons **(Cope)**

A moderate-sized (38–57 mm SVL) toad grayish to pale brown above (some yellow or green wash) with brown spots; vague light stripes on dorsum. Distribution (Fig. 117).

Figure 117

Figure 117   Distribution of *S. bombifrons*

4b Boss glandular .... **Great Basin Spadefoot,**
Scaphiopus intermontanus **(Cope)**

A moderate-sized (38–52 mm SVL) toad ash-gray above with indefinite olive area in center of dorsum. Distribution (Fig. 118).

Figure 118

Figure 118   Distribution of *S. intermontanus*

NEOTROPICAL FROG FAMILY
Leptodactylidae

The largest frog family, the Leptodactylidae, is composed of some 46 genera and 650–700 species distributed throughout the Neotropical Region. Several species occur in the southern parts of the United States. Some authorities would also include the Ghost Frogs (Heleophrynidae) of South Africa and the Australian bufonoids (Myobatrachidae) in the Leptodactylidae. Little evidence is available to link these groups together aside from past practices.

Leptodactylids are divided into four subfamilies, the largest of which (Telmatobiinae) is divided into six tribes. The family is exceptionally diverse, making prompt characterization difficult. Amplexus is inguinal in a few species, axillary in most. Most have external fertilization. Most have horizontal pupils and outer metatarsal tubercles but some southern genera have vertical pupils and lack the outer metatarsal tubercle. Most have classic arciferal pectoral girdles but one genus is secondarily firmisternal.

A diversity of reproductive biologies obtains in this family ranging from ovoposition (in foam nests or not) in ponds with typical pond-type larvae having horny beaks, labial denticles, and a sinistral spiracle to stream adapted larvae. Others have direct development (terrestrial eggs) and at least one gives birth to living young (*Eleutherodactylus jasperi*).

The most successful group of leptodactylids is the tribe Eleutherodactylini composed of some 13 genera. Eight of these are restricted to South America. *Sminthillus,* the smallest known frog (adult at 8–14 mm SVL), occurs in Cuba. *Hylactophryne, Syrrhophus,* and *Tomodactylus* primarily occur in Mexico. *Eleutherodactylus,* the largest vertebrate genus (450 species now recognized), occurs throughout the West Indies, South America north of Argentina, and in Central America to northern Mexico. The only species of *Eleutherodactylus* found in the United States were introduced into Fla.

1a Ventral surfaces of digit tips lacking discs (Fig. 119a).. **2**

1b Ventral surfaces of digit tips bearing discs (Fig. 119b).. **3**

a   b

Figure 119

Figure 119   Digit without (a) and with (b) discs

2a White stripe on upper lip; snout acuminate in dorsal view .... **White-lipped Frog,** *Leptodactylus fragilis* **(Brocchi)**

A moderate-sized (32–44 mm SVL) frog with a gray to brown ground color and dark brown spots; dorsolateral folds cream; white line on posterior surfaces of thighs; venter cream to white; extreme southern Tex. to Colombia and Venezuela.

2b Upper lip not striped; snout rounded in dorsal view ....
.... **Barking Frog,**
Hylactophryne augusti (Dugés)

4a Vomerine tooth patches in two broad fasiculi across palate (Fig. 121a) ....
.... **Greenhouse Frog,**
Eleutherodactylus planirostris (Cope)

Figure 120

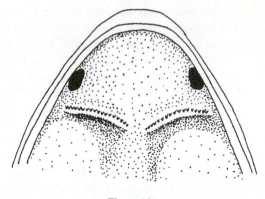

Figure 121a

Figure 120   *Hylactophryne augusti*

A moderate-sized (50–95 mm SVL) frog with smooth skin, dorsolateral folds, a transverse fold across occiput (Figure 120), prominent ventral disc. Brown above with darker blotches; venter cream. Central Tex., southern N. Mex., southern Ariz., thence south on Mexican Plateau to Oaxaca, Mexico.

Figure 121a   *Eleutherodactylus planirostris*

A minute (16–32 mm SVL) frog with a rusty-brown dorsum mottled with brown or bearing a pattern of stripes; digit tips bearing small pads; native to Cuba but widely introduced in the Caribbean and through peninsular Fla.

3a Vomerine odontophores posterior to choanae (Fig. 121) **4**

3b No vomerine odontophores on palate **5**

4b Vomerine tooth patches in two small clumps posterior and medial to choanae (Fig. 121B) .... Coqui Frog, *Eleutherodactylus coqui* **Thomas**

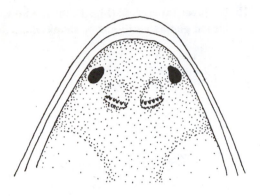

Figure 121b

Figure 121b   *Eleutherodactylus coqui*

A small (32–58 mm SVL) frog with a gray to brown dorsum mottled with darker pigment; venter cream to yellow finely stippled with brown pigment; digit tips bearing enlarged pads; native to Puerto Rico but introduced on St. Thomas and St. Croix (U.S. Virgin Islands) and at Fairchild Gardens, Miami, Fla.

5a Digit tips only slightly dilated; first finger longer than second .... Rio Grande Chirping Frog, *Syrrhophus cystignathoides* **(Cope)**

A minute (16–25 mm SVL) frog brown to olive above with tiny brown flecks; extreme southern Tex. to northern Veracruz, Mexico.

5b Digit tips obviously dilated; first finger equal in length to second 6

6a Dorsum flecked with brown .... Cliff Chirping Frog, *Syrrhophus marnockii* **Cope**

Figure 122

Figure 122   *Syrrhophus marnockii*

A small (19–38 mm SVL) frog with smooth skin; olive-green above with brown flecks (Fig. 122); south Central Tex.

6b Dorsum reticulated with brown .... Spotted Chirping Frog, *Syrrhophus guttilatus* **(Cope)**

A small (19–34 mm SVL) frog with smooth skin; brown above with a network of brown reticulation; Big Bend region of Tex. south to Guanajuato, Mexico.

TOAD FAMILY
Bufonidae

True toads (21 genera, 280 species) occur worldwide except in polar regions, the Pacific, Madagascar and the Australian Region, although the Marine Toad, *Bufo marinus,* is widely introduced. Three foci of generic endemism exist (tropical southeast Asia, Africa, and South America). *Bufo* ranges throughout the distribution of the family except in some high altitude situations.

Bufonids are edentate. Most have short limbs and are terrestrial. Large parotoid glands are present in all *Bufo* (and in some of the other genera). Most toads have thick warty skin. All have Bidder's Organ.

Amplexus is axillary (except in *Osornophryne*—inguinal) and fertilization is external except in *Nectophrynoides* (birth to living young). Tadpoles have horny beaks, labial teeth (2/3 rows), and a sinistral spiracle. Torrent-adapted tadpoles occur in the Asiatic *Ansonia* and the neotropical *Atelopus.*

Eighteen species of *Bufo* (including the introduced *B. marinus*) occur in the United States ranging in size from the diminutive Oak Toad (19–33 mm SVL) to the giant Marine Toad (100–240 mm SVL).

1a Inner tarsal fold present (Fig. 123a), leg glands in some 2

1b No inner tarsal fold (Fig. 123b) and no enlarged glands on thighs or shanks 6

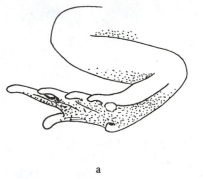

a

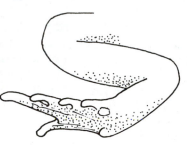

b

Figure 123

Figure 123   Inner tarsal fold present (a), absent (b)

2a Parotoid glands as large as side of head (Fig. 124); cranial crests low; no conspicuous glands on legs .... Giant Toad, *Bufo marinus* (Linnaeus)

Figure 124

Figure 124   *Bufo marinus*

A very large (100–240 mm SVL) toad brown above with indefinite black and pale brown flecks; obscure cranial crests; extreme southern Tex. to South America; introduced into vicinity of Miami.

2b Parotoid glands much smaller; conspicuous glands on legs 3

3a Cranial crests evident, curved around eye; skin relatively smooth (scattered warts) .... Colorado River Toad, *Bufo alvarius* Girard

Figure 125

Figure 125   *Bufo alvarius*

A large (75–150 mm SVL) toad olive or dark brown above; parotoid glands kidney-shaped; thigh and shank glands prominent (Fig. 125); southern Ariz. south to northern Sinaloa, Mexico.

3b Cranial crests lacking; skin warty 4

4a Parotoid glands large, flat; interval between parotoid glands less than width of gland .... Yosemite Toad, *Bufo canorus* Camp

A moderate-sized (45–71 mm SVL) toad sexually dimorphic in coloration. Males: yellow-green to olive above with scattered brown flecks. Females: ground color olive to yellow-brown with numerous dark brown to black blotches; central Sierra Nevada of Calif. (1850 to 3350 m).

4b Parotoid glands oval, separated by an interval at least twice width of a gland 5

5a Skin between warts smooth ....
.... **Black Toad,**
Bufo exsul **Myers**

A moderate-sized (47–71 mm SVL) toad with black flanks; dorsum olive-green to nearly black; warts pale colored; dorsal stripe cream; Deep Springs Valley, Inyo County, Calif.

5b Skin between warts tuberculate ....
.... **Western Toad,**
Bufo boreas **(Baird and Girard)**

A moderately large (50–120 mm SVL) toad with gray or greenish ground color and black spots; warts within spots rusty brown; white to yellow dorsal stripe. Distribution (Fig. 126).

Figure 126

Figure 126   Distribution of *B. boreas*

6a (1) Parotoid glands round ....
.... **Red-spotted toad,**
Bufo punctatus **Baird and Girard**

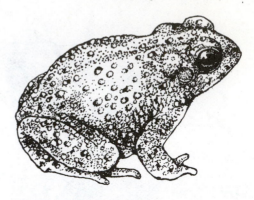

Figure 127

Figure 127   *Bufo punctatus*

A moderate-sized (38–76 mm SVL) toad gray to light brown above with buff or reddish warts; parotoid gland no larger than eye (Fig. 127); cranial crests absent or feebly developed. Distribution (Fig. 128).

Figure 128

Figure 128   Distribution of *B. punctatus*

6b Parotoid glands longer than wide........... 7

7a No cranial crests **8**

7b Cranial crests prominent **11**

8a Parotoid glands large, as long as side of head.. **9**

8b Parotoid glands much smaller.............. **10**

9a Dorsum reticulated with black .... **Sonoran Green Toad,** ***Bufo retiformis* Saunders and Smith**

A moderate-sized (38–58 mm SVL) toad boldly reticulated with black on green-yellow ground color; head and body flattened; south-central Ariz. to west-central Sonora, Mexico.

9b Dorsum flecked with black (Fig. 129) .... **Green Toad,** ***Bufo debilis* Girard**

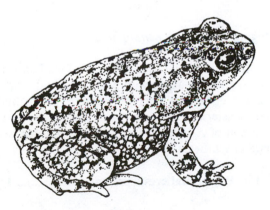

Figure 129

Figure 129   *Bufo debilis*

A moderate-sized (32–54 mm SVL) toad with black spots or bars on a bright green or yellow-

green ground color; head and body flattened. Distribution (Fig. 130).

Figure 130

Figure 130   Distribution of *B. debilis*

10a Anterior end of parotoid gland light-colored; metatarsal tubercles not sharp-edged .... **Southwestern Toad,** ***Bufo microscaphus* Cope**

A moderate-sized (51–76 mm SVL) toad; greenish-gray, brown, or buff above, buff below; throat not dark in males; isolated populations from southern Nev. and adjacent Utah across central Ariz. to western N. Mex. and southward into Mexico. Coastal southern Calif. to northern Baja California, Mexico.

10b Anterior end of parotoid not colored differently from rest of parotoid gland; inner metatarsal tubercle sharp-edged (sickle-shaped) .... Texas Toad, *Bufo speciosus* Girard

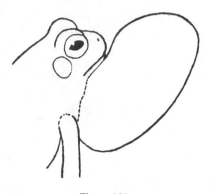

Figure 131

Figure 131   *Bufo speciosus*

A moderate sized (50–88 mm SVL) toad lacking a dorsal stripe; grayish above with uniform distribution of warts; parotoid gland oval; vocal sac of calling male sausage-shaped (Fig. 131); extreme southwestern Kans. and southeastern N. Mex. to Chihauhua, Coahuila, and Tamaulipas, Mexico.

11a Parotoid glands large, as long as head, adults minute .... Oak Toad, *Bufo quercicus* Holbrook

A minute (19–33 mm SVL) toad with a pale middorsal stripe bordered by 4–5 pairs of black or brown spots; Coastal Plain from Va. to eastern La., south through Fla.

11b Parotoid glands smaller; adults larger (40 mm SVL or more).................................. 12

12a Anterior one-half of parotoid glands pale colored .... see 10a

12b Anterior half of parotoid glands not colored differently from posterior half...... 13

13a No postorbital crests; cranial crests form boss between eyes (Fig. 132) .... Canadian Toad, *Bufo hemiophrys* Cope

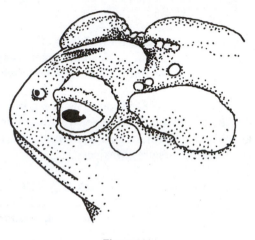

Figure 132

Figure 132   *Bufo hemiophrys*

A moderate-sized (50–83 mm SVL) toad brownish to greenish above with pale middorsal stripe; dark blotches on dorsum; western Minn. and northeastern S. Dak. northwestward to the eastern half of Alberta and southern Manitoba; isolate in southeast Wyo.

13b Postorbital crests present..................... 14

14a Parotoid gland triangular in shape; cranial crests elevated forming deep valley between them on top of head; prominent parietal crests.................. Gulf Coast Toad, *Bufo valliceps* Wiegmann

A moderately large (50–130 mm SVL) toad with a broad dark lateral stripe bordered above by a pale stripe; middorsal light stripe present; eastern La. to central Tex. south to Costa Rica.

14b Parotoid glands oval in outline; cranial crests lower; parietal crests not or only slightly developed 15

15a Parotoid glands strongly divergent; interorbital crests converge anteriorly forming boss on snout .... Great Plains Toad, *Bufo cognatus* **(Say)**

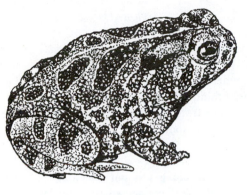

Figure 133

Figure 133   *Bufo cognatus*

A moderately large (48–115 mm SVL) toad gray or greenish above with large dark blotches edged with pale halos (Fig. 133); vocal sac sausage-shaped. Distribution (Fig. 134).

Figure 134

Figure 134   Distribution of *B. cognatus*

15b Parotoid glands not divergent; no snout boss 16

16a Interorbital crests ending in prominent knobs (Fig. 135)..
.. Southern Toad, *Bufo terrestris* **(Bonnaterre)**

Figure 135

Figure 135   *Bufo terrestris*

A moderately large (41–113 mm SVL) toad with high cranial crests; brown above with or without black spots (containing 1–2 warts). Distribution (Fig. 136).

Figure 136

Figure 136   Distribution of *B. terrestris*

16b Interorbital crests lacking knobs.......... 17

17a Warts on shank larger than those on thigh (Fig. 137)....................... American Toad,
Bufo americanus **Holbrook**

Figure 137

Figure 137 *Bufo americanus*

A moderately large (50–111 mm SVL) toad with only 1–2 warts per large dark spot (Fig. 137); parotoid gland kidney-shaped (Fig. 137), either not in contact with postorbital crests or connected via short spur; venter usually spotted with dark pigment. Distribution (Fig. 138).

Figure 138

Figure 138 Distribution of *B. americanus*

17b Warts on shank not larger than those on thigh....................................... 18

18a Parotoid gland touching postorbital ridge .... Woodhouse's Toad,
Bufo woodhousii **Girard**

Figure 139

Figure 139 Distribution of *B. woodhousii*

A moderately large (51–127 mm SVL) toad with a light middorsal stripe, prominent cranial crests, elongate parotoid glands, with an unmarked belly (often some spots on breast). Distribution (Fig. 139).

18b Parotoid gland not touching postorbital ridge but connected to it by lateral spur .... Houston Toad,
Bufo houstonensis **Saunders**

A moderate sized (51–79 mm SVL) toad with light middorsal stripe and mottled pattern of browns and blacks on cream to gray ground color; southeastern Tex.

TREE FROG FAMILY
Hylidae

This large family (36 genera, 570 species) is especially diverse in the neotropics. Hylids include those frogs with arciferal pectoral girdles having intercalary elements between the terminal and penultimate phalanges but not including Glass Frogs (family Centrolenidae) or Paradox Frogs (family Pseudidae). Although only two genera are currently recognized for the 125 species now reported from Australia and New Guinea, recently published work portends the recognition of several genera in that region.

Few species occur in Eurasia and only two dozen in North America north of Mexico. The major diversity occurs in the Neotropical Region where three very distinctive subfamilies co-occur with the more generalized Hylinae. Reproductive diversity is very apparent in the neotropics. Amplexus is axillary and fertilization is external. Eggs are deposited in a variety of situations (ponds, streams, vegetation over water, bromeliads, on the back of females, and in brood pouches on the female's back). Direct development occurs only in some of the forms depositing eggs on the back.

The 25 hylids found in the United States are best represented in the eastern part of the U.S.

1a Skin of head co-ossified with skull bones (not free) **2**

1b Skin of head not co-ossified with skull bones **3**

2a Pattern consisting of large brown blotches (Fig. 140)..
.... **Lowland Burrowing Treefrog,**
Pternohyla fodiens **Boulenger**

Figure 140

Figure 140   *Pternohyla fodiens*

A moderate sized (25–50 mm SVL) frog with prominent ridge between eye and nostril; small digital pads; webbing on toes reduced; one metatarsal tubercle; extreme south-central Ariz. along western Mexico to Michoacán.

2b Pattern consisting of indefinite small spots
or mottling ....
.... Cuban Treefrog,
Osteopilus septentrionalis (Boulenger)

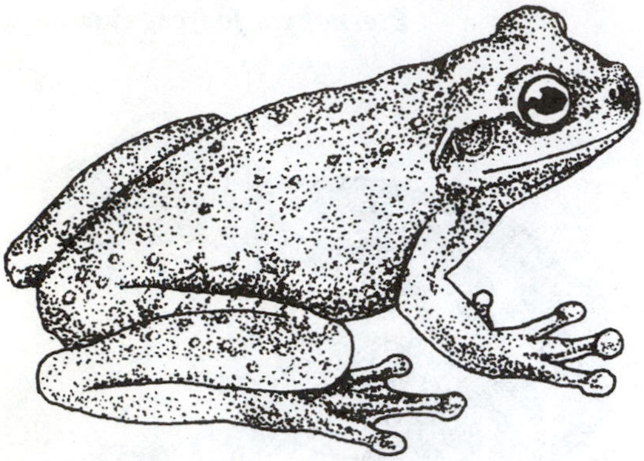

Figure 141

Figure 141 *Osteopilus septentrionalis*

A moderately large (38–140 mm SVL) frog with
large digital pads (Fig. 141); skin warty; brown
to dark olive above; southern Fla., Bahamas,
Cuba, and introduced in Puerto Rico.

3a Adults minute (less than 18 mm SVL) and
toes only ½ webbed (Fig. 142); digital pads
narrow........................ Little Grass Frog,
Limnaoedus ocularis (Holbrook)

Figure 142

Figure 142 Toe webbing in *Limnaoedus*

A minute (11–17 mm SVL) frog; dorsum brown,
olive, pink, or rust with tan dorsolateral stripe,
bordered below by brown stripe; venter white;
southeastern Va. to southwestern Ga. and eastern Fla. panhandle, south through Fla. peninsula.

3b Adults normally larger than 19 mm SVL
(if smaller, toes at least ⅔ webbed); digital
pads narrow to large................................ 4

4a Small digital pads present, scarcely wider than digit immediately below pad (Fig. 143a) **5**

4b Large digital pads present, twice as wide as digit immediately below pad (Fig. 143b) **13**

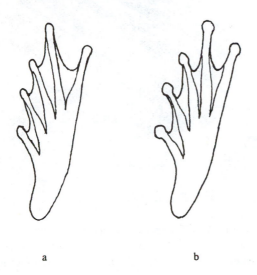

a   b

Figure 143

Figure 143   Small (a) and large (b) toe pads

5a 1½–3 joints of 4th toe free of webbing **6**

5b More than 3½ joints of 4th toe free of webbing **7**

6a Inner most toe webbed to disc; 4th toe with 1½–2 phalanges free of webbing ....
.... **Northern Cricket Frog,**
Acris crepitans **Baird**

Figure 144

Figure 144   *Acris crepitans*

A small (16–38 mm SVL) warty (Fig. 144) frog with gray to pale brown ground color and green, brown, or rust dorsal band; dark interorbital triangle present; dark stripe on posterior surfaces of thighs irregular laterally and blending dorsally with dark pigment on top of thighs. Distribution (Fig. 145).

Figure 145

Figure 145   Distribution of *A. crepitans*

6b Innermost toe ½ webbed; 3 phalanges of 4th toe free of web .... **Southern Cricket Frog,** *Acris gryllus* (LeConte)

A small (16–32 mm SVL) warty frog; dorsum light brown with darker brown markings; venter creamy yellow; posterior surfaces of thighs bearing clean-edged black stripe(s); Coastal Plain from southeastern Va. to Mississippi River.

7a Dorsal pattern consisting of three dark stripes or rows of dark spots **8**

7b Dorsal pattern consisting of two dark stripes (or rows of spots) or of irregular arrangements of spots **10**

8a Stripe on flanks darker than dorsal stripes .... **Brimley's Chorus Frog,** *Pseudacris brimleyi* Brandt and Walker

A small (25–32 mm SVL) frog with a tan ground color and three brown dorsal stripes; flank stripe black; chest spotted with brown; venter yellow; southeastern Va. to eastern Ga. on Coastal Plain.

8b Stripe or spots on flank same color as dorsal stripes or spots **9**

9a Lip stripe cream (not distinctly white) .... **Striped Chorus Frog,** *Pseudacris triseriata* (Wied)

Figure 146

Figure 146 *Pseudacris triseriata*

A small (19–38 mm SVL) frog with a brown interorbital triangle (or not) and brown dorsal stripes (or rows of spots) and lateral stripe (Fig. 146). Distribution (Fig. 147).

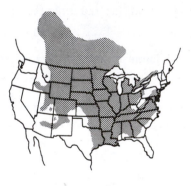

Figure 147

Figure 147 Distribution of *P. triseriata*

9b Lip stripe white (or lip bearing white spots) .... Southern Chorus Frog, *Pseudacris nigrita* (LeConte)

A small (19–32 mm SVL) frog with black markings on a pale gray background; snout more pointed than in other chorus frogs; Coastal Plain from N.C. to southern Miss. and south through the Fla. peninsula.

10a Bold black spots on flanks and in groin .... Ornate Chorus Frog, *Pseudacris ornata* (Holbrook)

A small (25–37 mm SVL) frog with bold black marks edged in cream on lateral surfaces; ground color green to brown, dorsal spots brown; Coastal Plain from N.C. to eastern La., south through much of Fla. peninsula.

10b Flanks and groin not so patterned 11

11a Black mark on upper lip below eye .... Strecker's Chorus Frog, *Pseudacris streckeri* Wright and Wright

A small (25–48 mm SVL) plump frog with spots on dorsum (may not be evident). Distribution (Fig. 148).

Figure 148

Figure 148   Distribution of *P. streckeri*

11b No black spot on upper lip 12

12a Dorsal stripes strongly curved (reverse parentheses) and nearly touching; stripes brown .... Mountain Chorus Frog, *Pseudacris brachyphona* (Cope)

A small (25–38 mm SVL) frog with an olive ground color and brown markings; western Pa. to central Miss.

12b Dorsal markings (spots normally) green, not forming reverse parentheses .... Spotted Chorus Frog, *Pseudacris clarki* (Baird)

A small (19–32 mm SVL) frog with green spots edged in black; ground color pale gray; venter white; central Kans. to west-central Tex. and extreme northern Tamaulipas, Mexico.

13a (4) Distal subarticular tubercles of toes not encompassed by toe webs .... Mountain Treefrog, *Hyla eximia* Baird

A small (18–57 mm SVL) green treefrog with short dorsal stripes (posterior half of body only) and complete flank stripes; central Ariz. (and adjacent N. Mex.) to Mexican Plateau.

13b Toe webbing reaching or encompassing distal subarticular tubercles 14

14a Pale spot below eye, usually between labial bars 15

14b No pale spot below eye, no labial bars 19

15a Tympanum nearly so large as eye; black patch on shoulder behind tympanum ....
.... Mexican Treefrog,
***Smilisca baudini* (Dumeril and Bibron)**

A moderately large (51–90 mm SVL) tree-frog gray, pale yellow, or brown above with dark brown blotches on dorsum; flanks mottled with brown; extreme southern Tex. through lowlands of Middle America to Costa Rica.

15b Tympanum distinctly smaller than eye 16

16a Canthal and supratympanic stripes evident; pattern of dorsum consists of large (larger than eye) brown blotches 17

16b No canthal or supratympanic stripes; pattern of dorsum consisting of small brown spots (smaller than eye) 18

17a Concealed surfaces of thighs washed with green or yellow-white ....
.... Bird-voiced Treefrog,
***Hyla avivoca* Viosca**

A moderate sized (29–52 mm SVL) treefrog; dorsum gray, brown, or green; extreme southern Ill. to eastern La. east to western Ga. and Fla.

17b Concealed surfaces of thighs washed with orange ....
.... Cope's Gray Treefrog,
***Hyla chrysoscelis* (Cope) and**
Gray Treefrog,
***Hyla versicolor* (LeConte)**

Figure 149

Figure 149   *Hyla chrysoscelis*

Slightly larger frogs (Fig. 149) (32–60 mm SVL); composite distribution (Fig. 150).

Figure 150

Figure 150   Distribution of *H. chrysoscelis* and *H. versicolor*

The two species may be distinguished by calls (fast trill—*H. chrysoscelis,* slow trill—*H. versicolor*) and karyotypes (*H. chrysoscelis* is diploid, *H. versicolor* is tetraploid).

18a Toe webbing reaching distal subarticular tubercle on outermost toe ....
.... **California Treefrog,**
Hyla cadaverina **Cope**

A moderate-sized (25–50 mm SVL) gray treefrog with brown spots; venter and concealed surfaces of limbs whitish; mountains of southern Calif. south to northern Baja California, Mexico.

18b Toe webbing encompassing proximal one-half of penultimate phalange of outermost toe .... Canyon Treefrog,
Hyla arenicolor **Cope**

A moderate-sized (32–57 mm SVL) gray or brownish treefrog blotched with brown; concealed surfaces of thighs orange-yellow. Distribution (Fig. 151).

Figure 151

Figure 151   Distribution of *H. arenicolor*

19a (15) Posterior surfaces of thighs brown or black with a row of pale spots ....
.... **Pine Woods Treefrog,**
Hyla femoralis **Latreille**

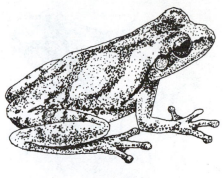

Figure 152

Figure 152   *Hyla femoralis*

A small (25–44 mm SVL) treefrog; dorsum reddish-brown to gray with or without brown blotches (Fig. 152); pale spots on thigh white, yellow, or orange; Coastal Plain from southeast Va. to eastern La. and south to southern Fla.

19b Posterior surfaces of thighs not spotted
.. 20

20a Prominent canthal and supratympanic stripes 21

20b Canthal and supratympanic stripes absent or ill-defined.. 22

21a Supratympanic stripe continuing as stripe along flank (edged above by white line) ....
.... **Pine Barrens Treefrog,**
Hyla andersoni **Baird**

A moderate-sized (29–51 mm SVL) green treefrog with lavender flank stripes edged with white; concealed surfaces of thighs orange; disjunct distribution—southern N.J., southeast N.C., southern Ga., Fla. panhandle.

21b Supratympanic stripe not continued as flank stripe... **Pacific Treefrog, *Hyla regilla* Baird and Girard**

22a Dorsum bearing X-shaped mark .... 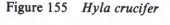 .... **Spring Peeper, *Hyla crucifer* Wied**

Figure 153

Figure 153 *Hyla regilla*

A small (18–50 mm SVL) treefrog; green or brown above; dark spots on dorsum (Fig. 153), or unicolor green; venter cream; concealed surfaces of thighs yellowish. Distribution (Fig. 154).

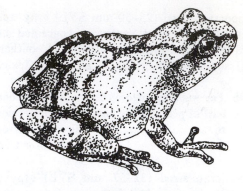

Figure 155

Figure 155 *Hyla crucifer*

A small (19–35 mm SVL) treefrog; dorsum brown, gray, or olive with dark brown X (Fig. 155). Distribution (Fig. 156).

Figure 154

Figure 154 Distribution of *H. regilla*

Figure 156

Figure 156 Distribution of *H. crucifer*

22b Dorsum variously marked, not bearing X-shaped mark **23**

23a Dorsum bearing numerous round dark brown spots..
.... **Barking Treefrog,**
Hyla gratiosa LeConte

A moderate-sized (51–70 mm SVL) treefrog; dark brown to bright green to pale gray spotted with dark brown; Coastal Plain from southeastern Va. to eastern La., south to southern Fla.

23b Dorsum not spotted, or if present, spots small and ill-defined.............................. **24**

24a A white or yellow stripe from below eye to groin..
.... **Green Treefrog,**
Hyla cinerea (Schneider)

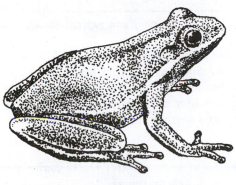

Figure 157

Figure 157   _Hyla cinerea_

A moderate sized (32–64 mm SVL) green tree-frog; lateral stripe of variable length (Fig. 157);

many specimens have minute gold spots on the dorsum. Distribution (Fig. 158).

Figure 158

Figure 158   Distribution of _H. cinerea_

24b No white lateral stripe ....
.... **Squirrel Treefrog,**
Hyla squirella Latreille

A small (22–41 mm SVL) brown (rarely green, temporary) treefrog; southeastern Va. to Fla., west to Corpus Cristi, Tex.

AQUATIC FROG FAMILY
Ranidae

This large family (43 genera, 600 species) is most prominent in Africa but is widely distributed except in Australia, the West Indies, and South America. *Rana* is the largest genus (250 species). Ranid frogs have firmisternal pectoral girdles, round sacral diapophyses, and usually have teeth. Most have smooth skin, long legs, and webbed feet. *Rana* ranges throughout the distribution of the family.

The brief synopsis given above is most inaccurate in Africa where fossorial, arboreal, and toad-like forms abound. Arboreal representatives likewise are abundant in the IndoAustralian archipelago.

Amplexus is axillary and fertilization external. In North America, larvae are of the pond-type with little variation except size, coloration, and numbers of tooth rows. Beaks are present and the spiracle is sinistral. Eggs are deposited as globular masses or as a surface film. Larval life ranges from two months to two years.

About two dozen forms are currently recognized in the United States. What was once "known" to be *Rana pipiens* has proven to be composed of many, mostly allopatric, species, not all of which have been named.

1a No cutaneous dorsolateral folds (see Fig. 159) 2

1b Distinct dorsolateral folds extending at least ⅔ length of body (see Figs. 161 and 165) 9

2a Ear distinct and large, at least ⅔ eye length in females, at least as large as eye in males; eastern species (Fig. 159) 3

2b Ear indistinct and small, about ½ eye length; western frogs 7

3a Adults small (41–76 mm SVL) 4

3b Adults larger (80–205 mm SVL) 5

4a Pale golden-brown dorsolateral stripes present ....
.... Carpenter Frog,
Rana virgatipes Cope

A moderate-sized (41–67 mm SVL) frog with 4 pale golden stripes (2 in dorsolateral positions, 2 lower on flanks) on a brown background; 2 phalanges of 4th toe free of web; Coastal Plain from southern N.J. to southern Ga.

4b Pattern variable, but no pale stripes ....
.... Mink Frog,
Rana septentrionalis Baird

A moderate-sized (48–76 mm SVL) brown frog spotted or mottled with dark brown; hind legs not cross-banded; 1–2 phalanges of 4th toe free of web; Labrador and northern Canada to northern Minn. and northern Great Lakes.

5a Light spots on lips; venter dark, spotted with pale markings ....
.... River Frog,
Rana heckscheri Wright

A large (80–135 mm SVL) brown to black frog (washed with green); tip of 4th toe free of web; southeastern N.C. to central Fla. and southern Miss.

5b Lips lacking white spots; venter pale with dark marbling 6

6a Tip of fourth toe free of web..... Bullfrog,
Rana catesbeiana Shaw

Figure 159

Figure 159   _Rana catesbeiana_

A large (80–205 mm SVL) green frog with or without brown mottling above. (Fig. 159). Distribution (Fig. 160).

Figure 160

Figure 160   Distribution of _R. catesbeiana_

6b Webbing reaching tip of fourth toe ....
.. Pig Frog,
Rana grylio Stejneger

A large (80–162 mm SVL) olive to dark brown frog spotted with black; venter white with brown marbling; southern S.C. to extreme southeastern Tex., south through Fla. peninsula.

7a (2) Throat normally spotted; markings on shank forming indistinct bars................. 8

7b Throat white to cream suffused with brown; shank bands distinct ....
.... Tarahumara Frog,
Rana tarahumarae Boulenger

A moderately large (62–102 mm SVL) frog olive to dark brown above with dark spots; toe tips bulbous; toes webbed to tips; extreme southern Ariz. south in Sierra Madre Occidental to Jalisco, Mexico.

8a Pale spot on snout; tips of toes black .... Mountain Yellow-legged Frog,
Rana muscosa Camp

A moderate-sized (50–81 mm SVL) brown frog with darker blotches; undersides of hind limbs yellow; toes webbed to tips; Sierra Nevada Mts of southern Calif.

8b No pale spot on snout; tips of toes not black .... Foothill Yellow-legged Frog,
Rana boyli Baird

A moderate-sized (44–68 mm SVL) brown frog with darker blotches; undersides of hind limbs and lower abdomen yellow; toes webbed to tips; west of Cascades in central Oreg. south to Los Angeles, Calif.

9a (1) Dorsolateral folds reaching sacral region (Figure 161) ....
.... **Green Frog,**
Rana clamitans **Latreille**

9b Dorsolateral folds reaching groin (see Fig. 168).. **10**

10a Webbing not penetrating beyond basal half or third of outer metatarsals of foot (Fig. 163a).. **11**

10b Webbing separating outer metatarsals of foot nearly to their base (Fig. 163b) **13**

11a 2.5–3 phalanges of 4th toe free of web; hind legs short; no dark face mask ....
.... **Crawfish Frog,**
Rana areolata **Baird and Girard**

Figure 161

Figure 161 *Rana clamitans*

A moderate-sized (54–102 mm SVL) frog green to bronze or brown above; venter white, spotted or marbled with dark brown; throat yellow in males; ear larger than eye in males (Fig. 161), about as large as eye in females; 1–2 phalanges of 4th toe free of web. Distribution (Fig. 162).

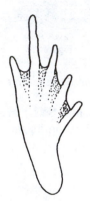

Figure 163a

Figure 163a *Rana areolata*

Figure 162

Figure 162   Distribution of *R. clamitans*

Figure 163b

Figure 163b   *R. cascadae*

A moderate-sized (57–108 mm SVL) frog cream to brown with large brown spots outlined with pale margins; venter white to heavily mottled; large lateral vocal sacs in males. Distribution (Fig. 164).

Figure 164

Figure 164   Distribution of *R. areolata*

11b 1–2 phalanges of 4th toe free of web; hind legs longer; a dark face mask present.. 12

12a Dark (black) face mask present (Fig. 165); about 2 phalanges of 4th toe free of web; vocal sacs in males .... Wood Frog, *Rana sylvatica* LeConte

Figure 165

Figure 165   *Rana sylvatica*

A moderate-sized (35–83 mm SVL) frog; pattern variable—ground color pink to brown to black; limbs banded; venter white. Distribution (Fig. 166). *Comment:* The population in Wyoming is sometimes recognized taxonomically as *Rana maslini* Porter.

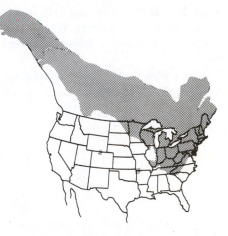

Figure 166

Figure 166   Distribution of *R. sylvatica*

12b Face mask dusky brown; 1–2 phalanges of 4th toe free from web; no vocal sacs in males .... Spotted Frog, *Rana pretiosa* Baird and Girard

A moderate-sized (50–100 mm SVL) brown frog with darker pale centered spots on the dorsum; venter yellow to red spotted or mottled with brown. Distribution (Fig. 167).

Figure 167

Figure 167   Distribution of *R. pretiosa*

13a Males lacking vocal sacs; dark facial mask usually evident; western Washington, Oregon, California, or Baja California 14

13b Males with vocal sacs; no facial mask; distributed east of above 15

14a Dorsum bearing black spots; abdomen and undersides of legs yellow .... Cascades Frog, *Rana cascadae* Slater

A moderate-sized (44–62 mm SVL) brown spotted frog; Cascade mountains from northern Washington to northern Calif.

14b Dorsum flecked with black; abdomen and undersides of legs yellow washed with red .... Red-Legged Frog, *Rana aurora* Baird and Girard

A moderately large (50–125 mm SVL) rust or brown frog with small black flecks on dorsum; venter spotted sparsely with dark brown; 1–2 phalanges of 4th toe free of web; southwestern British Columbia to northern Baja California west of Cascade-Sierran crest.

15a Dorsolateral folds not indented above groin; eastern Leopard frogs (Fig. 168a)........ 16

15b Dorsolateral folds indented above groin; western leopard frogs (Fig. 168b)......... 18

16a Dorsal pattern consisting of two rows of square spots between dorsolateral folds (Fig. 169).......................... Pickerel Frog, *Rana palustris* LeConte

Figure 169

Figure 169   *Rana palustris*

A moderate-sized (44–87 mm SVL) frog with tan to yellow-brown ground color and dark brown spots; groin and underside of legs yellow or orange; 2 phalanges of 4th toe free of web. Distribution (Fig. 170).

a

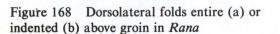
b

Figure 168

Figure 168   Dorsolateral folds entire (a) or indented (b) above groin in *Rana*

Figure 170

Figure 170   Distribution of *R. palustris*

16b Dorsal pattern consisting of round or elongate spots between dorsolateral folds... 17

17a No white spot on eardrum; often spot on top of snout; flanks boldly spotted ....
.... Northern Leopard Frog,
***Rana pipiens* Schreber**

Figure 172

Figure 172   Distribution of *R. pipiens*

17b White spot on eardrum; no snout spot; flanks with few spots ....
.... Southern Leopard Frog,
***Rana utricularia* Harlan**

A moderate-sized (51–127 mm SVL) green or brown frog with elongated spots. Distribution (Fig. 173).

Figure 171

Figure 171   *Rana pipiens*

A moderate-sized (51–111 mm SVL) green or brown frog with elongate spots (often margined with pale borders) between dorsolateral folds (Fig. 171). Distribution (Fig. 172).

Figure 173

Figure 173   Distribution of *R. utricularia*

The western Leopard Frogs remain taxonomically unstable. Named forms include *Rana berlandieri* Baird (Rio Grande Leopard Frog), *R. blairi* Mecham *et al* (Plains Leopard Frog), *R. chiricahuensis* Platz and Mecham (Chiricahua Leopard Frog), and possibly the poorly-known *Rana fisheri* Stejneger and *R. onca* Cope. An undescribed species ("lowland type") found in Arizona is also part of the complex. *Rana fisheri* and *R. onca* do not have medially indented dorsolateral folds and may be part of the *pretiosa* complex.

 Rana berlandieri is a pale leopard frog with Mullerian ducts in the male; it normally lacks a snout spot. It occurs from central Texas and southern N. Mex. south into Mexico.

 Rana blairi is a tan leopard frog with numerous small round spots on the dorsum and normally has a snout spot (Fig. 174). It lacks Mullerian ducts in males. It occurs from western Ind. to eastern Colo. and then south through the Plains to central Tex. (see Fig. 175).

 Rana chiricahuensis is a brown leopard frog with or without Mullerian ducts in the male; the venter is heavily suffused with gray. The posterior surfaces of the thighs are brown with small cream spots. It occurs in central Ariz. east to western N. Mex. and in southeastern Ariz. (and adjacent N. Mex.) south to southern Chihuahua, Mexico.

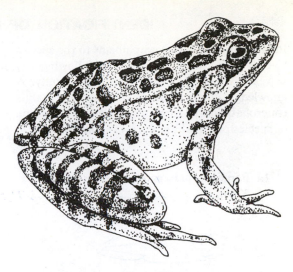

Figure 174

Figure 174   *Rana blairi*

Figure 175

Figure 175   Distribution of *R. blairi*

IDENTIFICATION OF LARVAL AMPHIBIANS

Identification of larval amphibians to the species level is not always possible. The following key serves to identify larval forms as to family. A knowledge of distribution and breeding habits (microhabitat and time) may allow identification of species of larval amphibians in many cases.

1a Forelimbs present, no adhesive organs (Fig. 176) **2**

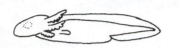

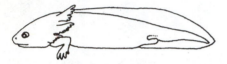

Figure 176   Developmental stages of a salamander

1b Forelimbs absent, adhesive organ is present (Fig. 3, p. 5) **13**

2a Gills *or* gill slits present **3**

2b Both external gills and open gill slits present **5**

3a External gills present (*Notopthalmus, Taricha*) .... (part) Salamandridae

3b External gills not present, gill slits present **4**

4a Tongue present, costal grooves 16 or fewer (postlarval *Cryptobranchus*) .... (part) Cryptobranchidae

4b Tongue absent, costal grooves 55 or more .... (part) Amphiumidae

5a Hind limbs absent .... **Sirenidae**

5b Hind limbs present **6**

6a Digits 3–3 or 2–2, 55 or more costal grooves .... (part) Amphiumidae

6b Digits 4–4 or 4–5, 24 or fewer costal grooves **7**

7a 2 gill slits .... **Proteidae**

7b 3–4 gill slits **8**

8a 3 gill slits (larval and neotenic *Eurycea,*
Gyrinophilus, Haideotriton, Hemidacty-
lium, Pseudotriton, Typhlomolge, Ty-
phlotriton, Stereochilus) .... (part)
Plethodontidae

8b 4 gill slits **9**

9a Prevomerine teeth in single series, palatine
teeth present or absent but, if present, in
single series.. **10**

9b Prevomerine teeth and palatine teeth pres-
ent and in patches................................ **12**

10a Palatine teeth present, maxillary teeth ab-
sent (*Desmognathus* and *Leurognathus*)
.... (part) **Plethodontidae**

10b Palatine teeth absent, maxillary teeth
present **11**

11a First gill smallest (larval *Cryptobranchus*)
.... (part) **Cryptobranchidae**

11b First gill largest (*Dicamptodon* and
Rhyacotriton)..
.... (part) **Ambystomatidae and**
Dicamptodontidae

12a Maxillary teeth absent ....
.... (part) **Salamandridae**

12b Maxillary teeth present ....
.... (part) **Ambystomatidae**

13a (1) Mouth bearing horny beaks and at least
4 rows of denticles (Fig. 177)............... **16**

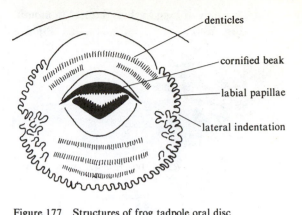

Figure 177   Structures of frog tadpole oral disc

13b Mouth lacking horny beaks and denticles
.... **14**

14a Tadpole bearing barbels about the mouth;
two lateral spiracles **15**

14b Tadpole lacking barbels; one ventral spir-
acle (Figs. 178, 179) ....
.... **Microhylidae**

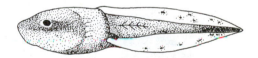

Figure 178   Microhylid tadpole

Figure 179   Mouth of *Microhyla* larva

15a One pair of long barbels (lateral to mouth) (Fig. 180).......................... Pipidae

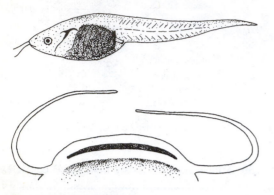

Figure 180 Pipid larva

15b Four pair of short barbels (Fig. 181) .... Rhinophrynidae

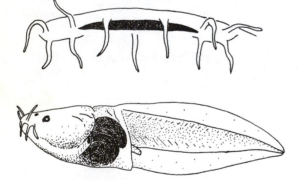

Figure 181 Rhinophrynid larva

16a Spiracle single, under throat; tooth rows 2–3/10–12 (Fig. 182) .... Ascaphidae

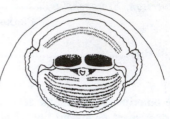

Figure 182 Ascaphid larva

16b Spiracle single, sinistral (on left side of body); tooth rows usually 2/3 but as great as 7/6 17

17a Papillary border complete above oral disc or having a very narrow gap; tooth formula at least 2/4 (Fig. 183) .... Pelobatidae

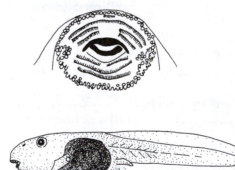

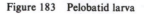

Figure 183 Pelobatid larva

17b **Papillary border broadly interrupted above oral disc (gap about equal to length of upper tooth rows)** **18**

18a **Papillary border incomplete below oral disc (Fig. 184)** .... **Bufonidae**

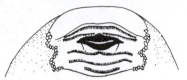

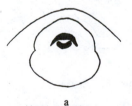

Figure 184   Bufonid larva

18b **Papillary border complete below oral disc** **19**

19a **Anus median (Fig. 185a)** (*Leptodactylus fragilis*) .... **Leptodactylidae**

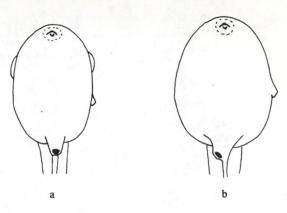

Figure 185   Anal opening medial (a) and dextral (b)

19b **Anus dextral (Fig. 185b)** **20**

20a **Oral disc emarginate (Fig. 186a)** .... **Ranidae**

20b **Oral disc not emarginate (Fig. 186b)** .... **Hylidae**

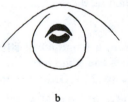

Figure 186   Oral disc emarginate (a) or not (b)

Key to Orders and Suborders[3] of Reptiles

1a Body form "rounded" with bony or leathery shell (turtles) (Fig. 187) ....
.... (p. 95) Order Testudinia

Figure 187   Turtle

1b Body form elongate, no shell 2

2a Cloacal aperture, a longitudinal slit (Fig. 188a) (crocodiles and alligators) ....
.... (p. 213) Order Crocodylia

2b Cloacal aperture, a transverse slit (Fig. 188b) Order Squamata ....
.... 3

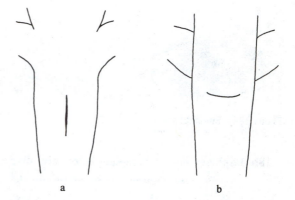

a   b

Figure 188   Cloacal slit of crocodiles (a) and squamates (b)

3a No legs, ear openings nor eyelids (Fig. 189) (snakes) ....
.... (p. 165) Suborder Serpentes

Figure 189   Snake

3b Two pairs of legs, or it not, with eyelids and usually ear openings (Fig. 190) (lizards) .... (p. 122) Suborder Lacertilia

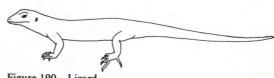

Figure 190   Lizard

[3]Also known by ordinal names derived from Greek, i.e. Chelonia=Testudinia, Ophidia=Serpentes, Sauria=Lacertilia.

TURTLES,
Testudinia (Chelonia)

Salamanders may be confused with lizards and limbless lizards are often mistaken for snakes but the body form of turtles is universally recognizable. The ancient reptilian subclass Anapsida survives today as the Order Testudinia, whose body form has not changed since their postulated origin in late Paleozoic. Of the two suborders of modern turtles, only the Cryptodira is represented in North America. The Pleurodira or side-necked turtles (so named because the neck is retracted under the shell by bending it sideways) are restricted to Australia, Africa, and South America, although fossil pleurodires are known from North America. Cryptodires, which retract their head by bending the neck vertebrae in an S-shaped curve, are represented on all continents, although only by marine forms along Australian coasts.

The most unique feature of turtles is their shell which is composed of bony plates covered by horny shields. The shell consists of a dorsal carapace and a ventral plastron. The number and shapes of the epidermal shields are typically used for identification (Fig. 191). Other important aids of identification include color patterns and type of feet which are modified according to type of locomotion being elephant-like in terrestrial tortoises, flipper shaped in marine turtles or variable but with webbed digits in aquatic species.

Turtles have an anapsid skull with no temporal opening and usually strong muscular necks and jaws. There are no teeth but rather both upper and lower jaws form a horny beak which is sharp and can inflict considerable damage. Most turtles are omnivorous, eating a variety of plant and animal material. Many turtles are carnivorous as young, feeding on insects, worms and other small animals and are increasingly herbivorous as adults. The Green Turtle and gopher tortoises are strictly vegetarians.

Activity of turtles in temperate climates is usually limited to the warm months from approximately April to October but sea turtles and

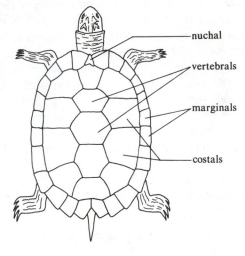

Figure 191a

Figure 191a   Shields of carapace

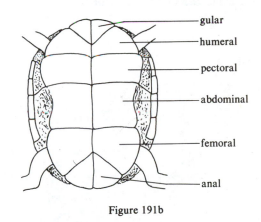

Figure 191b

Figure 191b   Shields of plastron

a few species in the southeastern U.S. are active all year. Most turtles are diurnal but some aquatic forms (musk turtles, mud turtles and snapping turtles) are more active at night. Aestivation is common in aquatic and semiaquatic turtles during drought periods and hibernation is typical during cold winter months.

Turtles are noted for their relatively slow growth and long lives. Growth of species in temperate climates is recorded by the concentric rings formed on the epidermal scutes as each is enlarged at its margin. Box turtles are known to live in excess of 100 years and the famous Galapagos tortoises off the South American coast may live for 200 years or more. As a group, turtles are probably the longest lived vertebrates. Their slow growth and typically delayed maturation make them particularly susceptible to population decline as they are easily collected and exploited by man for sport, to be eaten, or taken as pets. Certain aquatic turtles are known to transmit *Salmonella* (a bacterium which causes food poisoning) to man and some states have banned such species from the pet market.

Reproduction occurs in the spring with mating typically in April and May. All species are oviparous and lay relatively large eggs which may be soft- or hard-shelled. Some species may lay several clutches in one season. Nesting usually occurs in June or July and hatching follows in August and September. There is no parental care and nests are not guarded.

Of the approximately 330 species of world turtles, 48 species occurring in the continental U.S. and surrounding oceans represent 7 families.

KEY TO THE FAMILIES OF TURTLES

1a Carapace covered with horny plates....... 3

1b Carapace not covered with horny plates but is smooth and leathery or fleshlike......... 2

2a Limbs paddle-like (Fig. 192a), no claws, snout not tubular, longitudinal ridges on carapace and plastron, marine .... (p. 98) LEATHERBACK TURTLE FAMILY, Dermochelyidae

2b Limbs not paddle-like, feet webbed with claws (Fig. 192b), snout tubular, body flattened, no longitudinal ridges, not marine .... (p. 98) SOFT-SHELLED TURTLE FAMILY, Trionychidae

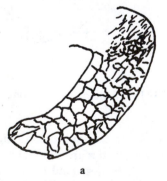
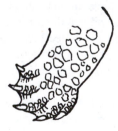

a   b

Figure 192   Turtle limbs

3a Plastron reduced, small and cruciform (Fig. 193), head conspicuously large and tail long usually more than half the length of the carapace... (p. 100) .... SNAPPING TURTLE FAMILY, Chelydridae

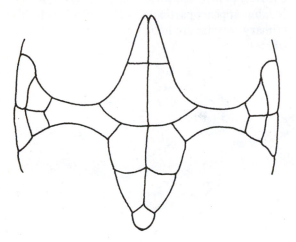

Figure 193   Plastron of Chelydrids

3b Plastron not reduced, head not noticeably enlarged, and tail not long....................... 4

4a Hind feet elephant-like (Fig. 194) .... (p. 101) LAND TORTOISE FAMILY, Testudinidae

Figure 194   Hindfoot of land tortoises

4b Hind feet not elephant-like...................... 5

5a Limbs paddle-like (Fig. 192a), strictly marine.. .... (p. 103) SEA TURTLE FAMILY, Cheloniidae

5b Limbs not paddle-like (Fig. 192b) 6

6a Plastron composed of 12 scutes; pectoral scutes form part of bridge with carapace and contact marginals (Fig. 195a)............. .... (p. 109) BOX, POND, and MARSH TURTLE FAMILY, Emydidae

6b Plastron composed of 10 or 11 horny scutes; pectoral scutes not forming part of bridge with carapace and do not contact marginals (Fig. 195b).................... (p. 105) MUSK and MUD TURTLE FAMILY, Kinosternidae

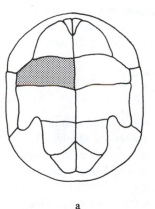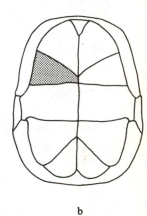

a   b

Figure 195   Plastron, pectoral scutes forming (a) or not forming (b) part of bridge

LEATHERBACK TURTLE FAMILY
Dermochelyidae

The largest turtle in the world is the leatherback turtle which is the only living member of the family Dermochelyidae. The leatherback, a strictly marine species, may measure as much as 70 inches long and weigh 1500 pounds. The leatherback, *Dermochelys coriacea* Linnaeus, which is divided into the Atlantic leatherback (*D. c. coriacea*) and the Pacific leatherback (*D. c. schlegeli*) is wide ranging in tropical and subtropical oceans around the world. In the U.S. it is known to nest on both coasts of Florida. Except for nesting, it is entirely pelagic.

The common name derives from the fact that adults lack horny plates typical of turtles. The carapace and plastron are covered with a tough leathery skin and have five prominent ridges in addition to ridges at the margin of the carapace (Fig. 196). Hatchlings have small scales on the skin which are gradually lost with age. The shell bones are also lost except for the nuchal and marginals of the plastron and have been replaced by a mosaic of small bony platelets embedded in the skin. Color of the carapace is brown to black and the plastron is white. The head and limbs may have white or yellow blotches along the sides.

The limbs are highly modified into flippers for its aquatic existence. The leatherback lays 50–170 eggs and shares its nesting grounds with other marine turtles. It is omnivorous, feeding on a variety of invertebrates, fish, and floating seaweed, but appears particularly fond of jellyfish. Its primary enemies are man, large sharks, and killer whales.

Figure 196

Figure 196   *Dermochelys coriacea*

SOFT-SHELLED TURTLE FAMILY
Trionychidae

The soft-shelled turtles (Fig. 197) are so named because they lack horny epidermal shields. The carapace is covered with a soft leathery skin. They have long necks and flat shell and occur entirely in freshwater. There are approximately 22 species in Africa, Asia, and North America of which 3 species occur in the United States.

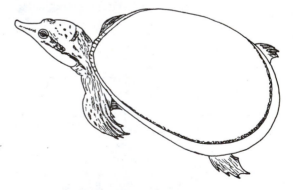

Figure 197

Figure 197   *Trionyx*

1a Carapace smooth, without tubercles, nostrils round with no ridge projecting from central septum (Fig. 198a)........................ .... Smooth Softshell, *Trionyx muticus* LeSueur

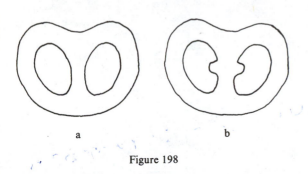

Figure 198

Figure 198   Nostrils without (a) and with (b) ridge on central septum

Olive to orange-brown, adults with pattern of darker blotches, Adults 11–35 cm shell length. Distribution (Fig. 199).

Figure 199

Figure 199   Distribution of *Trionyx muticus*

1b Carapace with tubercles, nostrils with ridge projecting from central septum (Fig. 198b) .... 2 .

2a Tubercles blunt knobs particularly concentrated along anterior edge and marginal ridge .... Florida Softshell *Trionyx ferox* Schneider

Color gray to brown may have darker blotches; adults 15–46 cm, Restricted to Fla. peninsula and portions of southeastern Ala., southern Ga., and southwestern S.C.

2b Tubercles numerous and conical giving carapace a sandpaper-like surface .... Spiny Softshell, *Trionyx spiniferus* LeSueur

Color olive to tan, pattern of dark blotches or circles and a marginal dark line; Adults 13–46 cm. Distribution (Fig. 200).

Figure 200

Figure 200   Distribution of *T. spiniferus*

SNAPPING TURTLE FAMILY
Chelydridae

Chelydridae is a new world family consisting of three species which occur in the United States. Snapping turtles are among the largest freshwater turtles characterized by very large head, hook-shaped beak, and powerful jaws. The plastron is greatly reduced and there is a long tail.

1a **A row of scutes (supramarginals) between the marginals and the first three pleurals (Fig. 201), upper jaw strongly hooked, carapace with three prominent keels ....** .... **Alligator Snapping Turtle,** *Macroclemys temminckii* **(Troost)**

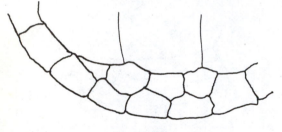

Figure 201

Figure 201   Supramarginals of *Macroclemys*

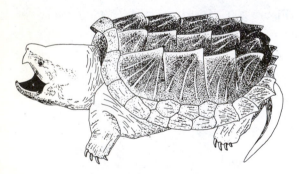

Figure 202

Figure 202   *Macroclemys temminckii*

Carapace dark, plastron grayish (Fig. 202). Adults 38–66 cm, 35–150+ pounds. Distribution (Fig. 203).

Figure 203

Figure 203   Distribution of *M. temminckii*

1b **No supramarginals, upper jaw not strongly hooked, carapace not strongly keeled 2**

2a **Width of third vertebral the same or greater than height of second pleural ....** .... **Florida Snapping Turtle,** *Chelydra osceola* **Stejneger**

Carapace olive-brown to dark brown; tail with prominent keel; adults 20–47 cm; restricted to Peninsular Fla.

2b Width of third vertebral much less than height of second pleural.............................
.... Snapping Turtle,
Chelydra serpentina (Linnaeus)

Carapace brown or olive to black; tail with prominent keel (Fig. 204); adults 20–47 cm. Distribution (Fig. 205).

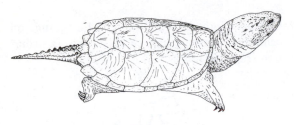

Figure 204

Figure 204 _Chelydra serpentina_

Figure 205

Figure 205   Distribution of _C. serpentina_

LAND TORTOISE FAMILY
Testudinidae

The 10 genera and 39 living species of generally large land tortoises are found on all continents except Australia and also occur on some oceanic islands. They are strictly terrestrial and largely vegetarian. Growth rings on scutes are prominent. There are three species of a single genus in the U.S.

1a Head wedge-shaped when viewed from above (Fig. 206a), carapace length less than 2 times its maximum height.......................
... Texas Tortoise,
Gopherus berlandieri (Agassiz)

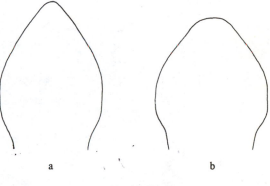

a   b

Figure 206

Figure 206   Head shape outlines of _G. berlandieri_ (a) and other _Gopherus_ (b)

Figure 207

Figure 207   *Gopherus berlandieri*

Carapace brown, scutes with yellow centers, adults 10–22 cm (Fig. 207), restricted to extreme Southern Tex. and adjacent northeastern Mexico.

1b Head rounded in front when viewed from above (Figure 206b), carapace flatter, its length more than 2 times its maximum height 2

2a Distance from base of first and fourth claw on forefoot equal to distance between first and fourth claw on hindfoot .... Desert Tortoise, *Gopherus agassizi* (Cooper)

Carapace light brown with yellow or orange on centers of scutes; carapace 15–37 cm. Distribution (Fig. 208).

Figure 208

Figure 208   Distribution of *G. agassizi*

2b Distance from base of first and fourth claw on forefoot greater than same distance on hindfoot.......................... Gopher Tortoise, *Gopherus polyphemus* (Daudin)

Figure 209

Figure 209   Distribution of *G. polyphemus*

Carapace brown or brownish yellow, scutes with lighter centers, carapace 15–37 cm; restricted to southeastern U.S. (Fig. 209).

SEA TURTLE FAMILY
Cheloniidae

The sea turtles include 4 genera of strictly marine turtles with streamlined body forms and large paddle-like forelimbs. The neck is short and the head cannot be retracted into the shell. All species are pelagic and return to land only to nest. Of the 6 species occurring in the warmer waters of the tropic and subtropical parts of the world, 5 species occur in the coastal waters of the U.S.

1a One pair of prefrontal scales on head (Fig. 210a); lower jaw strongly serrate .... Green Turtle, *Chelonia mydas* **(Linnaeus)**

Figure 211

Figure 211 *Chelonia mydas*

Carapace olive to brown (Fig. 211), occasionally with pattern of radiating lines; carapace 90–140 cm; name derives from greenish body fat; Pacific and Atlantic coastal waters.

1b Two pairs of prefrontal scales on head (Fig. 210b); lower jaw not strongly serrate..... 2

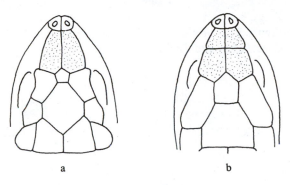

a   b

Figure 210

Figure 210 Prefrontals of *Chelonia mydas* (a) and other (b) sea turtles

2a Four pleurals; first pleural does not touch nuchal; large carapace scutes overlap ....
.... **Hawksbill,**
Eretmochelys imbricata (Linnaeus)

Figure 212

Figure 212 *Eretmochelys imbricata*

Carapace 76–90 cm, brown, smaller individuals with "tortoise shell" pattern of lighter markings (Fig. 212); Atlantic coast of U.S., Pacific coast of Baja but occasional specimens may stray to southern Calif. coast.

2b Five or more pleurals, first pleural touches nuchal, large carapace scutes do not overlap **3**

3a Three enlarged inframarginals at bridge (Fig. 213a)...
.... **Loggerhead,**
Caretta caretta (Linnaeus)

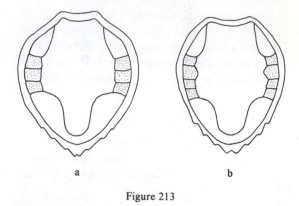

a   b

Figure 213

Figure 213 Enlarged inframarginals in *Caretta* (a) and *Lepidochelys* (b)

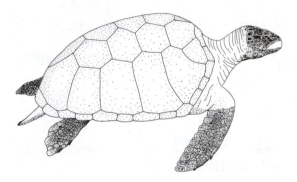

Figure 214

Figure 214 *Caretta caretta*

Carapace reddish-brown, 79–120 cm (Fig. 214); Atlantic coast, Pacific coast to southern Calif.

3b Four enlarged inframarginals at bridge (Fig. 213b)... **4**

4a Carapace gray, Atlantic coast .... Atlantic Ridley, *Lepidochelys kempi* (Garman)

4b Carapace olive green, Pacific coast........... .... Pacific Ridley, *Lepidochelys olivacea* (Eschscholtz)

Carapace 60–71 cm; rare along Calif. coast.

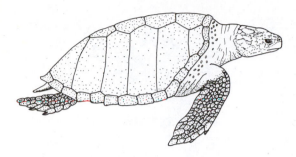

Figure 215

Figure 215   *Lepidochelys kempi*

Carapace 40–70 cm (Fig. 215), distributed in Gulf stream to Newfoundland.

MUSK AND MUD TURTLE FAMILY
Kinosternidae

Kinosternidae is a New World family of small aquatic, bottom-dwelling turtles related to the snapping turtles. In the older literature they are considered a subfamily of Chelydridae. They have only 10–11 plastral scutes and possess musk glands associated with the bridge which emit an odorific secretion when disturbed. There are four genera and 25 species. Two genera and 8 species occur in the U.S.

1a Single indistinct hinge along anterior edge of abdominal scute; length of interfemoral suture equal or greater than interhumeral suture (Fig. 216a) (*Sternotherus*) 2

1b Two distinct transverse hinges bordering anterior and posterior margins of abdominal scutes; length of interfemoral suture much less than interhumeral suture (Fig. 216b) (*Kinosternon*)................................ 4

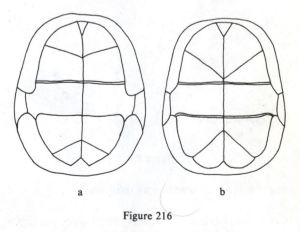

a   b

Figure 216

Figure 216   Plastron of *Sternotherus* (a) and *Kinosternon* (b)

2a Shields of carapace not overlapping (Fig. 217a), barbels on throat and chin.............. .... Stinkpot,
Sternotherus odoratus (Latreille)

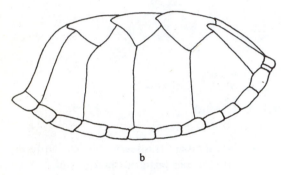

a

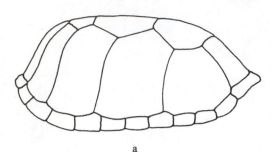

b

Figure 217

Figure 217   Carapace shields not overlapping (a) and overlapping (b)

Figure 218

Figure 218   *Sternotherus odoratus*

Color grayish brown to black, no distinct pattern in adults, juveniles with pattern of dark spot or

radiating lines; two prominent light (yellow) lines on each side of head and neck (Fig. 218); size 8–12 cm. Distribution (Fig. 219).

Figure 219

Figure 219   Distribution of *S. odoratus*

2b Shields of carapace overlapping (Fig. 217b), barbels on chin only............................... 3

3a Gular scute present, vertebral keel not prominent (may also possess lateral keels) (Fig. 220a)....... Loggerhead Musk Turtle,
Sternotherus minor (Agassiz)

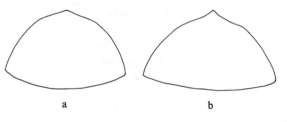

a   b

Figure 220

Figure 220   Vertebral keel weak (a) and prominent (b) [in cross section]

Color tan to brown, darker lines on head, neck and carapace; size 7–12 cm. Distribution (Fig. 221).

Figure 221

Figure 223

Figure 221   Distribution of *Sternotherus minor*

Figure 223   Distribution of *S. carinatus*

3b Gular scute absent, vertebral keel well developed (Fig. 220b)............................. .... Razorback Musk Turtle, *Sternotherus carinatus* (Gray)

4a Carapace with three longitudinal light lines........................... Striped Mud Turtle, *Kinosternon bauri* Garman

Figure 222

Figure 224

Figure 222   *Sternotherus carinatus*

Color brown, dark markings on carapace (Fig. 222) which tend to disappear with age, head with small dark spots; size 10–13 cm. Distribution (Fig. 223).

Figure 224   *Kinosternon bauri*

Color brown to olive with 3 light stripes on carapace (Fig. 224); size 7–10 cm; Key West and peninsular Fla. to extreme southern Ga.

4b Carapace without longitudinal light lines 5

5a Ninth marginal noticeably higher than eighth (Fig. 225a).. **Yellow Mud Turtle,** *Kinosternon flavescens* (Agassiz)

5b Ninth marginal not greatly higher than eighth (equal or only slightly higher) (Fig. 225b) **6**

6a First vertebral in contact with second marginal (Fig. 228a).. **7**

6b First vertebral not in contact with second marginal (Fig. 228b) .... **Eastern Mud Turtle,** *Kinosternon subrubrum* (Lacépède)

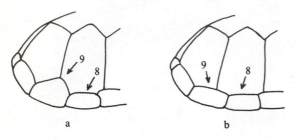

Figure 225

Figure 225 Marginal scutes in *Kinosternon*

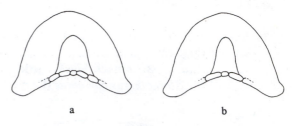

Figure 228

Figure 228 First vertebral in contact (a) or not in contact (b) with second marginal

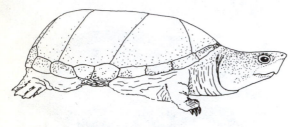

Figure 226

Figure 226 *Kinosternon flavescens*

Color brown to olive or green, chin and throat yellowish (Fig. 226); size 10–14 cm. Distribution (Fig. 227).

Carapace dark brown, unmarked; head may be marked with light lines or spots; size 7–12 cm. Distribution (Fig. 229).

Figure 227

Figure 227 Distribution of *K. flavescens*

Figure 229

Figure 229 Distribution of *K. subrubrum*

7a Shell flattened, twice as wide as high in cross section (Fig. 230a) .... Sonoran Mud Turtle, *Kinosternon sonoriense* LeConte

7b Shell arched, not flattened, less than twice as wide as high in cross section (Fig. 230b) .... Mexican Mud Turtle, *Kinosternon hirtipes* Wagler

Carapace olive-brown to black, head with small dark spots or mottling; size 10–17 cm; Texas Big Bend and south into Mexico.

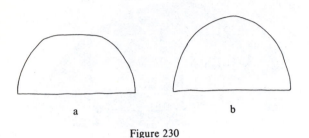

a   b

Figure 230

Figure 230   Carapace shape in Sonoran (a) and Mexican (b) Mud Turtles

Carapace brown, head with mottled dark pattern; size 11–16 cm; southeastern Ariz. to western Tex. and south into Mexico.

BOX, POND, AND MARSH TURTLE FAMILY
Emydidae

The emydids comprise the largest family of turtles with more than 80 species in 25 genera in North and South America, Europe, Asia, and Africa. The 26 species and 7 genera in the United States comprise a representative cross-section of the family which includes species which are entirely terrestrial (e.g., box turtles) as well as those which are principally aquatic. Males of pond turtles frequently have much elongated claws on the forelimbs.

2a Upper jaw notched at anterior tip (Fig. 231a) .... Blanding's Turtle, *Emydoidea blandingi* (Holbrook)

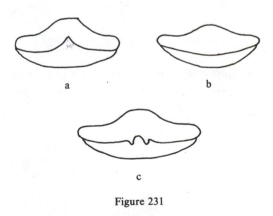

a   b

c

Figure 231

Figure 231   Anterior views of jaws

1a Plastron with well developed hinge(s) 2

1b Plastron without well developed hinge ... 3

Figure 232

Figure 232   *Emydoidea blandingi*

Carapace dark brown to black with yellow flecks or spots (Fig. 232), chin yellow; size 12–20 cm. Distribution (Fig. 233).

Figure 233

Figure 233   Distribution of *E. blandingi*

2b Upper jaw not notched at tip (Fig. 231b)
··· 7

3a Neck very long approximately equal to plastron length .... Chicken Turtle, *Deirochelys reticularia* (Latreille)

Figure 234

Figure 234   *Deirochelys reticularia*

Carapace dark brown with light yellow netlike pattern (Fig. 234); long neck has yellow stripes; size 10–20 cm. Distribution (Fig. 235).

Figure 235

Figure 235   Distribution of *D. reticularia*

3b Neck short, about one-half plastron length
··· 4

4a Upper jaw notched and bordered by tooth-like projections (Fig. 231c)...................... 8

4b Upper jaw not notched 5

5a Crushing surface of upper jaw is narrow 14

5b Crushing surface of upper jaw is broad 6

6a Carapacial scutes roughened by concentric growth rings, neck not striped .... Diamondback Terrapin, *Malaclemys terrapin* (Schoepff)

Figure 236

Figure 236   *Malaclemys terrapin*

Carapace with concentric dark rings on each scute (Fig. 236) over ground color of gray, orange or yellow; head and neck with small dark spots; size 10–20 cm. Distribution in costal marshes, tidal flats, and estuaries from southern Tex. to Cape Cod, Mass.

6b Carapacial scutes smooth, neck striped 17

7a (2) Carapace with vertebral keel (ridge) .... Eastern Box Turtle, *Terrapene carolina* (Linnaeus)

Figure 237

Figure 237   *Terrapene carolina*

Color pattern highly variable; carapace high, dome shaped (Fig. 237); size 10–17 cm. Distribution (Fig. 238).

Figure 238

Figure 238   Distribution of *T. carolina*

7b Carapace without vertebral keel ....
.... **Western Box Turtle,**
Terrapene ornata (Agassiz)

**8a Posterior margin of carapace not serrated;
scutes of carapace smooth** ....
.... **Painted Turtle,**
Chrysemys picta (Schneider)

Figure 239

Figure 239   *Terrapene ornata*

Carapace dark brown to black with radiating yellow lines (Fig. 239); size 10–15 cm. Distribution (Fig. 240).

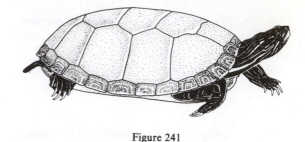

Figure 241

Figure 241   *Chrysemys picta*

Carapace brown with variable patterns of red, yellow or olive markings depending on subspecies; head and neck striped yellow (Fig. 241); size 10–20 cm. Distribution (Fig. 242).

Figure 240

Figure 240   Distribution of *T. ornata*

Figure 242

Figure 242   Distribution of *C. picta*

**8b Posterior margin of carapace serrated;
scutes of carapace wrinkled with furrows**
.... **9**

9a Prominent patch(es) of red or yellow on side of head .... Slider, *Chrysemys scripta* (Schoepff)

10a Second costal scute with light C-shaped marking (Fig. 245)... River Cooter, *Chrysemys concinna* (LeConte)

Figure 243

Figure 243 *Chrysemys scripta*

Carapace pattern variable, olive with dark markings or dark brown to black with yellow markings (Fig. 243); size 12–25 cm. Distribution (Fig. 244).

Figure 245

Figure 245 *Chrysemys concinna*

Color pattern variable, generally with concentric alternating dark and light markings; prominent light stripes on neck; plastron with numerous dark markings; size 18–40 cm. Distribution (Fig. 246).

Figure 244

Figure 244 Distribution of *C. scripta*

9b Lines but no patches on side of head ... 10

Figure 246

Figure 246 Distribution of *C. concinna*

10b Second costal scute not marked as above 11

11a Plastron red, orange, or coral **12**

11b Plastron not red, orange or coral.............. .... **Cooter,** *Chrysemys floridana* **(LeConte)**

Figure 247

Figure 247   *Chrysemys floridana*

Color pattern of irregular dark lines (Fig. 247) on lighter background (*C.f. floridana)* or light lines on dark background *(C.f. peninsularis);* size 23–40 cm. Distribution (Fig. 248).

Figure 248

Figure 248   Distribution of *C. floridana*

12a Stripes on side of neck end behind eyes.... .... **Florida Redbelly Turtle,** *Chrysemys nelsoni* **(Carr)**

Figure 249

Figure 249   *Chrysemys nelsoni*

Carapace dark brown to black with red or yellow markings on pleura (Fig. 249), plastron reddish; size 20–30 cm; lower three quarters of Fla.

12b Stripes on side of neck continue between eyes and onto snout **13**

13a Carapace arched, highest near center, restricted to Atlantic Coastal Plain .... **Red Belly Turtle,** *Chrysemys rubriventris* **(Babcock)**

Coloration variable but usually dark with red and/or light markings on carapace, plastron light with variable dark blotches and bordered with red; size 25–35 cm; along mid Atlantic states from N.J. to N.C. and west into W. Va.

13b Carapace flattened, restricted to Gulf coastal plain .... **Alabama Redbelly Turtle,** *Chrysemys alabamensis* **(Baur)**

Carapace dark green to black with red to yellow bars and reddish plastron; size 20–25 cm; occurs in salt marshes in southern Ala. and extreme western Fla.

14a (5) Scutes of carapace sculptured with ridges .... **Wood Turtle,** *Clemmys insculpta* **(LeConte)**

Figure 250

Figure 250   *Clemmys insculpta*

Carapace brown with black and yellow lines radiating from centers of pleural shields (Fig. 250), orange on neck and limbs; size 12–20 cm. Distribution (Fig. 251).

Figure 251

Figure 251   Distribution of *C. insculpta*

14b Scutes of carapace smooth not sculptured **15**

15a Carapace dark with distinct small scattered yellow spots (Fig. 252) .... **Spotted Turtle,** *Clemmys guttata* **(Schneider)**

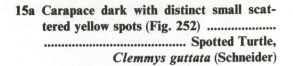
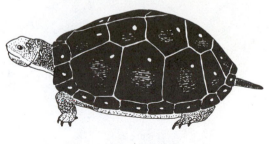

Figure 252

Figure 252   *Clemmys guttata*

Carapace bluish-black with yellow spots; size 8–12 cm. Distribution (Fig. 253).

Figure 253

Figure 253   Distribution of *C. guttata*

15b Carapace without distinct spots **16**

16a Prominent, large orange patch on head (Fig. 254) .... Bog Turtle, *Clemmys muhlenbergi* (Schoepff)

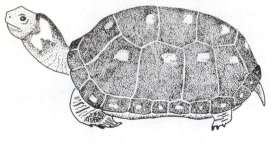

Figure 254

Figure 254   *Clemmys muhlenbergi*

Carapace dark brown with yellowish or reddish centers of scutes; size 8–11 cm. Distribution (Fig. 255).

Figure 255

Figure 255   Distribution of *C. muhlenbergi*

16b No orange patch on head .... Western Pond Turtle, *Clemmys marmorata* (Baird & Girard)

Carapace dark, (olive, brown or black) with brown or black lines, often broken and radiating from center of scutes; size 9–18 cm; along Pacific coast from northern Baja California through Oreg. and disjunctly into Wash.

17a (6) Carapace with well developed keel forming a middorsal ridge with prominent projections 18

17b Carapace without well developed keel of prominent projections 24

18a Projections of middorsal keel blunt, rounded and black (Fig. 256)...................... .... Black-knobbed Map Turtle, *Graptemys nigrinoda* Cagle

Figure 256

Figure 256   *Graptemys nigrinoda*

Brownish carapace with fine light rings on shields, neck greenish with yellow lines; size 7–18 cm; Restricted to Alabama, Tombigbee, and Black Warrior River systems in Ala. and Miss.

18b Projections of middorsal keel not blunt and knob-like 19

19a A large light (orange or yellow) spot on each pleural shield... .... Yellow-blotched Map Turtle, *Graptemys flavimaculata* Cagle

Carapace brown with distinct orange to yellow spot on scutes, mid-dorsal keel prominent; size 7–18 cm; restricted to southeastern Miss.

19b No distinct light spots as above 20

20a Pleural shields with light ring or oval mark .... **Ringed Map Turtle,** *Graptemys oculifera* **Baur**

Carapace brown with distinct rings of red to yellow color on pleural shields, prominent keels darker; size 7–22 cm; southern Miss. and La. in Pearl River system.

20b Not as above **21**

21a Distinct thin yellow crescent behind eye separates neck stripes from eye (Fig. 257) .... **Mississippi Map Turtle,** *Graptemys kohni* **Baur**

Figure 257

Figure 257   *Graptemys kohni*

Carapace brown with complex pattern of lighter lines; size 9–25 cm. Distribution (Fig. 258).

Figure 258

Figure 258   Distribution of *G. kohni*

21b Yellow marks behind eye not as above **22**

22a Large solid light blotch or band behind eye separates neck stripes from eye (Fig. 259) **23**

Figure 259

Figure 259   *Graptemys pulchra*

22b Light blotch or band behind eye variable but smaller and does not prevent neck stripe from reaching eye .... **False Map Turtle,** *Graptemys pseudogeographica* **Gray**

Carapace brownish to olive, with yellow irregular lines on pleura and along seams of marginals; size 9–27 cm. Distribution (Fig. 260).

Figure 260

Figure 260   Distribution of *G. pseudogeographica*

23a Longitudinal light bar under chin (Fig. 261a) .... **Alabama Map Turtle,**
Graptemys pulchra Baur

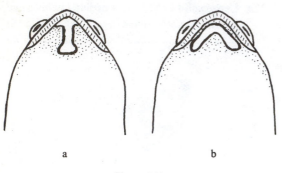

a   b

Figure 261

Figure 261   *G. pulchra* (a), *G. barbouri* (b)

Figure 262

Figure 262   *Graptemys pulchra*

Carapace brown to green, marginals with semi-circular orange or yellowish light marks (Fig. 262); size 9–30 cm. Distributed in southern Miss., southern Ala., extreme eastern La., and extreme western Fla. in Escambia, Alabama, and Pearl River systems.

23b Curved light bar under chin (Fig. 261b) .... **Barbour's Map Turtle,**
Graptemys barbouri Carr & Marchand

Carapace dark brown to olive with light oval or semicircular markings on pleura and marginals becoming obscured in old individuals; size 9–32 cm; western panhandle of Fla. (Apalachicola River system) and into southern Ala. and Ga.

24a (17) A longitudinal light mark behind eye .... **Map Turtle,**
Graptemys geographica (LeSueur)

Carapace olive to brown with reticulate pattern of light marks; size 10–28 cm. Distribution (Fig. 263).

Figure 263

Figure 263   Distribution of *Graptemys geographica*

24b Light mark behind eye not as above 25

25a Curved transverse light mark under chin (Fig. 264a) .... Cagle's Map Turtle, *Graptemys caglei* Haynes & McKown

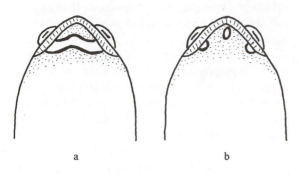

a   b

Figure 264

Figure 264   *G. caglei* (a), *G. versa* (b)

Carapace green with irregular light marks, light V-shaped mark on top of head with arms forming crescent behind eyes; size 7–16 cm; restricted to south central Tex. in Guadalupe and San Antonio Rivers.

25b Central oval spot under chin with single lateral circular spots posteriorly on each side of chin (Fig. 264b) .... Texas Map Turtle, *Graptemys versa* Stejneger

Carapace olive with irregular light marks; size 7–13 cm; J-shaped orange to red mark behind eyes; Colorado River drainage in central Tex.

SQUAMATE REPTILES
Squamata

Historically, the Squamata has been divided into the lizards and the snakes, but students of the "ringlizards" (amphisbaenians) have argued that these burrowers constitute a third coordinate group (= groups of equal rank) of squamates. If the Squamata is considered an order, the Amphisbaenia, Lacertilia, and Serpentes are each considered suborders. If those three groups are called orders, then the Squamata must be a superorder. We here continue the practice of using (as biological groups) these groups, although they are illogical under the theory of evolution and the present hypothesis of relationships (e.g., Lacertilia), because these group names and their status (as coordinate groups) are traditionally employed to categorize organisms and are embedded for the present in the language of biology and everyday use and because we are not prepared to propose here a reclassification of amphibians and reptiles. The more serious student, in need of more than identification, should refer to the following account (otherwise proceed to Amphisbaenia, Lacertilia, or Serpentes).

PROBLEMS WITH THE CLASSIFICATION OF SQUAMATES

Most biologists have assumed that each of the groups they recognized is monophyletic in the sense of Simpson (= minimal monophyly) where all members of the group are postulated to be derived from a common ancestor, but have failed to appreciate that monophyly must include a second component (that all descendants of that common ancestor be members of the group as well), otherwise, *any* assembly of organisms becomes monophyletic (see Farris, 1974).

The difficulties of discovery of sufficient evidence to support these hypotheses of strict monophyly are considerable because the Squamata is a group peculiar within the tetrapods for its propensity towards limb loss and the concomitant adjustments to burrowing. Snakes, ring-lizards, and several obscure lizard groups parallel one another by the losses of fields of structures involving the paired appendages. Bilateral symmetry is disrupted with reduction or loss of paired organs as the organisms become elongate. The fusions and/or losses of skull elements commonly seen within the Squamata are interpreted as adaptations to burrowing. Once attention is focused upon the trends toward limb loss, simplification of the skull, and disruption of symmetry, it is not surprising that a good deal of emphasis was placed on retention of primitive features (e.g., traces of limbs in snakes, which organ is not reduced, "normal" arrangement of cranial bones, etc.).

The Squamata is a very varied group of Reptiles. Nevertheless, the group is considered monophyletic because it embraces the only organisms known having hemipenes (paired eversible intromittant organs). Every species of Squamata (except those comprised only of females) has these retractable organs in the base of the tail. Squamates can also be characterized as Lepidosaurians having lost the lower temporal cranial arch, thereby, allowing movement of the quadrate (not true of *Sphenodon punctatus,* the only living Rhynchocephalian).

Either of these traits is sufficient to advance the hypothesis of monophyly of the Squamata but neither is sufficient to advance a particular hypothesis of the interrelations among or the monophyly of various subgroupings of squamates (i.e., lizards, ring-lizards, and snakes). The usual diagnosis of the lizards is one emphasizing the postulated general (= primitive) conditions of squamates (having limbs, eyelids, ears, etc.), whereas, those for ring-lizards and snakes emphasize the necessary losses of some of those primitive conditions.

If a subgroup is characterized by a particular feature, it does not follow that the remaining subgroups (not having the feature) form an exclusive group. For example, ring-lizards have an enlarged, median premaxillary tooth (sufficient to postulate a hypothesis of monophyly for ring-lizards), but the absence of such a feature is insufficient to postulate the hypothesis that snakes and lizards form a second monophyletic unit of squamates.

Most authors agree that ring-lizards comprise a monophyletic group, that snakes comprise a monophyletic group, and that each group is derived from ancestors having suites of features consistent with their classification as lizards. "Lizards", thus, equals "non-snake" and "non-ring."

If lizards, ring-lizards, and snakes are each monophyletic and coordinate (= equal rank), one must represent this hypothesis in the form of a non-testable hypothesis (Fig. 265a). Three restrictive (= testable) hypotheses are possible under this non-testable one: (i) lizards and snakes share a common ancestor not shared with ring-lizards (Fig. 265b), (ii) lizards and ring-lizards share a unique common ancestry (Fig. 265c), and (iii) ring-lizards and snakes share a unique common ancestry (Fig. 265d). Most herpetologists accept another hypothesis (Fig. 265e), i.e., that snakes and ring-lizards are independently derived from lizards. This hypothesis may well be true, but if true, Amphisbaenia, Lacertilia, and Serpentes cannot be classified as coordinate groups, because some "lizards" are more closely related to snakes than they are to other lizards (i.e., the classification is inconsistent with the hypothesis of relationships and with the observations used to generate the hypothesis).

The discussion given above is intended to alert the serious student of a philosophical debate within biology. There is a growing tendency for evolutionary biologists to demand that classifications must be consistent with the theories of evolution and inheritance if the classification is to have any relevant information content and if classification can serve as a basis for generating testable hypotheses and predictions. However,

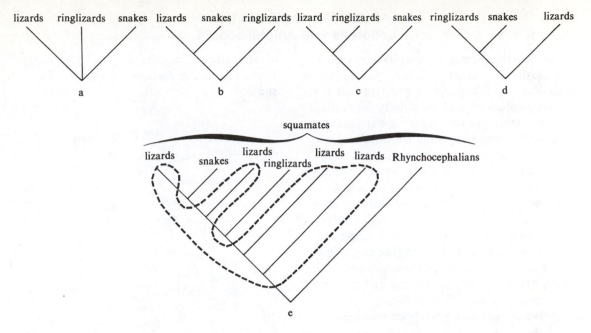

Figure 265

Figure 265   Cladograms of squamate relationships (see text for explanation)

these exercises may appear counter-intuitive. After all, species 1 and species 2 will be species 1 and species 2 no matter what the classification. They will look the same and carry out their biologies regardless of whether or not the classification employed is consistent with the theory of evolution. However, a classification based on overall similarities will not allow a scientist to study the pattern produced by the process of evolution.

REFERENCES

Farris, J. S. 1974. Formal definitions of paraphyly and polyphyly. Systematic Zoology 23:548–554.

Gans, C. 1978. The Characteristics and affinities of the Amphisbaenia. Trans. Zool. Soc. London 34:347–416.

Mayr, E. 1981. Biological classification: toward a synthesis of opposing methodologies. Science 214:510–516.

McDowell, S. B., Jr., and C. M. Bogert. 1954. The systematic position of *Lanthanotus* and the affinities of the anguinomorphan lizards. Bull. Amer. Mus. Nat. Hist. 105:1–142.

Platnick, N. I. 1977. Paraphyletic and polyphyletic groups. Systematic Zoology 26:195–200.

Underwood, G. 1967. A contribution to the classification of snakes. Publication No. 653, x + 179 pp. British Museum (Natural History).

Wiley, E. O. 1975. Karl R. Popper, systematics, and classification—a reply to Walter Bock and other evolutionary taxonomists. Systematic Zoology 24:233–243.

————. 1981a. Phylogenetics/The theory and practice of phylogenetic systematics. xv + 439 pp. Wiley-Interscience.

————. 1981b. Convex groups and consistent classifications. Systematic Botany 6:346–358.

LIZARDS, SQUAMATA
Lacertilia and Amphisbaenia

Lizards, like their reptilian "cousins" the snakes, are typically unpopular with most people. Unfortunately, too few people appreciate that lizards are inoffensive and beautifully interesting creatures. Most lizards feed on insects and are beneficial in consuming large numbers of pests. All lizards will retreat rather than attack an approaching person, but most will permit close observation of their varied and interesting behaviors. While most lizards will bite in self defense upon being caught, their small conelike teeth preclude serious injury to people and only the larger species are even capable of "breaking the skin". Only two lizards in the world are poisonous and both of these (Gila Monster of Arizona and adjacent regions and Beaded Lizard of Mexico) are non-aggressive, secretive and rarely encountered.

Like other reptiles, lizards are "cold-blooded" (poikilothermic) and require heat from the environment to maintain body temperatures sufficiently high for normal activity. Partly because of this fact lizards are commonly seen basking and many are easily observed. More species occur in warmer climates where they comprise a sizable part of the entire fauna. None are known in the arctic or alpine tundra of North America and only a few species occur as far north as Canada. The lizard fauna of the deserts of the southwestern United States is particularly large and variable where a majority of the 100 species and eight of the nine families of U.S. lizards may be found.

Lizards possess typical reptilian characteristics and usually may be distinguished from snakes by the presence of legs. However, legs are absent in the glass lizards *(Ophisaurus),* the legless lizard *(Anniella pulchra),* and the worm lizard *(Rhineura floridana).* These legless lizards may be distinguished from snakes by the presence of movable eyelids (which incidentally are absent in some geckos and in Xantusiidae) and by the presence of external ear openings (except

in *Anniella, Cophosaurus* and *Holbrookia).* Typical structural features are shown in Fig. 266 and 267.

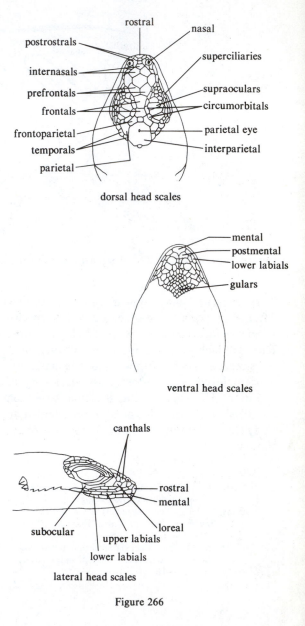

dorsal head scales

ventral head scales

lateral head scales

Figure 266

Figure 266   Head scales of lizards

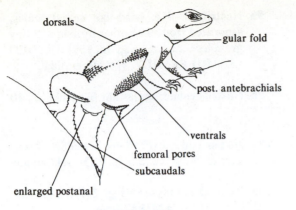

Figure 267

Figure 267   Typical characteristics of lizards

KEY TO THE FAMILIES OF LIZARDS AND AMPHISBAENIANS

1a Two pair of legs 4

1b Legs absent 2

2a Eyes absent .... (p. 158) FLORIDA WORM LIZARD FAMILY, Rhineuridae

2b Eyes present 3

3a Ear opening present .... (p. 153) ALLIGATOR LIZARD FAMILY, Anguidae

3b No ear opening .... (p. 156) LEGLESS LIZARD FAMILY, Anniellidae

4a Eyelids absent 5

4b Eyelids present 6

5a Head covered with large flat plates .... (p. 157) NIGHT LIZARD FAMILY, Xantusiidae

5b Head covered with small granular scales .... (p. 125) GECKO FAMILY, Gekkonidae

6a Fold of skin along side of body between limbs containing granular scales much smaller than those above or below the fold .... (p. 153) Anguidae

6b No such fold of skin 7

7a Belly scales larger than dorsal scales, quadrangular and in longitudinal series (Fig. 268).................. (p. 159) WHIPTAIL FAMILY, Teiidae

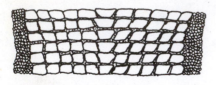

Figure 268

Figure 268   Ventral scales of Teiidae

7b Belly scales smaller or not in a longitudinal series 8

8a Dorsal scales smooth, shiny and flat, approximately equal in size with overlapping rounded borders .... (p. 146) SKINK FAMILY, Scincidae

8b Dorsal scales granular or keeled (possessing a ridge), not smooth, shiny and flat.. 9

9a Nostril slit-like; bead-like dorsal scales, separated from each other by skin and tiny granules .... (p. 159) BEADED LIZARD FAMILY, Helodermatidae

9b Nostrils round; scales not bead-like 10

10a No parietal eye (third eye in middle of rear part of skull) .... (p. 125) Gekkonidae

10b Parietal eye present (Fig. 269) ..... (p. 129) IGUANID FAMILY, Iguanidae

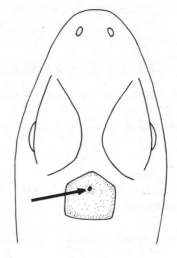

Figure 269

Figure 269   Parietal eye of Iguanidae

GECKO FAMILY
Gekkonidae

Geckos comprise a family of generally small, soft skinned lizards. Although there are relatively few species in the United States, the family is large (approximately 75 genera and 600 species) and cosmopolitan with many species occurring in tropical parts of the world. Four subfamilies (some authors recognize these as distinct families) which have been described include the Eublepharinae (*Anarbylus* and *Coleonyx* in the U.S.) with true eyelids and no friction pads, the Gekkoninae (*Hemidactylus* and *Phyllodactylus* in the U.S.) with a transparent spectacle covering the eye and highly modified digital pads, the Sphaerodactylinae (*Sphaerodactylus* and *Gonatodes* in the U.S.) with no eye covering and variously modified digital pads, and the Diplodactylinae of Australia and New Zealand.

Geckos are typically nocturnal with well developed eyes, although the Sphaerodactylinae tend to be diurnal. Except for the ground geckos (Eublepharinae), most geckos have conspicuous toe pads which are used to provide friction foot holds for climbing. Geckos are often transported in fruit and materials from different parts of the world and are likewise often associated with human habitation, frequently living in and around dwellings. The Mediterranean Gecko is a good example of a lizard whose natural distribution has been greatly expanded by man's activities. It has been introduced widely from its native Mediterranean-African distribution. Geckos are insectivorous and are, therefore, useful companions of man. The tongue is thick and not extensible. Some geckos have reasonably loud voices and can make a squeaking or barking noise. Geckos are oviparous, producing only one or two eggs at a time, although many species, particularly in the tropics, may reproduce continuously throughout the year.

1a Eyelids present 2

1b Eyelids absent 5

2a Dorsal scales uniformly small and granular 4

2b Dorsal scales of enlarged tubercles and smaller granules....................... 3

3a Enlarged transverse lamellae on toes .... **Reticulated Gecko,** ***Coleonyx reticulatus*** **Davis & Dixon**

Light tan with profusion of dark spots and reticulations; total length 14–17 cm; restricted to small area of Big Bend of Tex.

3b Enlarged transverse lamellae absent .... **Switak's Barefoot Gecko,** ***Anarbylus switaki*** **Murphy**

Gray to brown with transverse rows of yellow spots or bands across dorsum; body size 5–8 cm; eastern San Diego and southwestern Imperial Counties in Calif., ranging southward into Baja California.

4a Preanal pores (4–6) separated at midline by 1 or more scales (Fig. 270a)................
.... **Texas Banded Gecko,**
Coleonyx brevis **Steneger**

4b Preanal Pores (6–10) not separated at midline (Fig. 270b).............................
.... **Western Banded Gecko,**
Coleonyx variegatus **(Baird)**

Dark brown bands over light yellow to pink background color with bands irregular to blotched and variegated pattern in adults; body 5–7.5 cm. Distribution (Fig. 272).

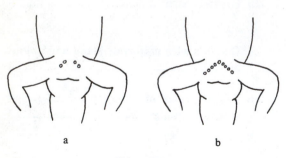

a   b

Figure 270

Figure 270   Preanal pores, *C. brevis* (a); *C. variegatus* (b)

Figure 271

Figure 271   *Coleonyx brevis*

Color consisting of dark brown transverse bars wider than light yellow background color (Fig. 271). Bars breaking up to brown spots or patches in older adults; Body 4.5–6 cm.; southern N. Mex., southwestern Tex. and north central Mexico.

Figure 272

Figure 272   Distribution of *C. variegatus*

5a Toes without expanded pads ....
.... **Yellow-Headed Gecko,**
Gonatodes albogularis
(Dumeril & Bibron)

Head and neck of males yellow to orange. Body uniform gray to brown or blotched irregularly with dark and light spots of gray and brown, tip of tail white unless regenerated; body small to 4 cm.; introduced in southern Fla. and Keys.

5b Toes with expanded pads at least at tips
.... **6**

6a A single rounded scale pad at tip of toes (Fig. 273), dorsal scales uniform in size.... 7

Figure 273

Figure 273   Toe of *Sphaerodactylus*

6b Not a single round scale at tip of toe, but rather a larger pad on toe at tip or not, dorsal scales granular but not uniform in size (Fig. 275) 9

7a Dorsal scales granular and smaller than ventral scales 8

7b Dorsal scales larger than ventral scales .... Reef Gecko, *Sphaerodactylus notatus* Baird

Figure 274

Figure 274   *Sphaerodactylus notatus*

Color of small dark spots on yellowish-brown background (Fig. 274). Some individuals with three darkened lines on head and neck; females often with two white spots on shoulders; body size to 3 cm; Florida Keys and extreme southeastern Fla.

8a Light stripes on head and white spots on neck .... Ocellated Gecko, *Sphaerodactylus argus* Gosse

Brown to olive brown, patternless or with light spots, tail brown to reddish brown; body size to 3 cm; introduced into Key West.

8b No light stripes or white spots .... Ashy Gecko, *Sphaerodactylus cinereus* Wagler

Color reddish-brown to gray with numerous small yellow spots scattered over surface; body size to 4 cm; introduced into Florida Keys.

9a Toe pads consisting of two series of enlarged scales at tips of toes separated by claw (Fig. 275a) .... Leaf-toed Gecko, *Phyllodactylus xanti* Cope

9b Toe pads consisting of series of enlarged scales in middle of toe with small distal tip (Fig. 275b) .... Mediterranean Gecko, *Hemidactylus turcicus* (Linnaeus)

Figure 277

Figure 277   *Hemidactylus turcicus*

Color white to tan with dark irregular spots on dorsal surface (Fig. 277); body size to 6.5 cm; introduced species now well established in gulf coastal states.

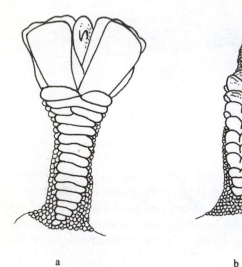

a   b

Figure 275

Figure 275   Toes of *Phyllodactylus* (a) and *Hemidactylus* (b)

Figure 276

Figure 276   *Phyllodactylus xanti*

Color light gray to yellowish with irregular dark gray splotches and bands (Fig. 276); body size to 6 cm; southern Calif. south through Baja California.

IGUANID FAMILY
Iguanidae

The Iguanidae is the largest and most diverse family of lizards in the western hemisphere as well as in the United States. Lizards of this family are also those species which are most conspicuous and easily observed. Their diverse body forms reflect the wide range of habitats in which they may be found. One of their most distinct characteristics is the nature of their tooth attachment being on the inside of the tooth ridge (pleurodont) rather than on top. The tongue is thick and not extensible. Their size varies from the small spindly aboreal anoles to the large terrestrial chuckwalla (larger iguanids occur outside the U.S., including the rock iguana, *Cyclura,* of the Caribbean Islands, the Galapagos land, *Conolophus,* and marine, *Amblyrhynchus,* iguanas and the green iguana, *Iguana iguana,* of Central America). Most are insectivorous, but some of the large species are almost exclusively vegetarian. One of the most interesting features of the iguanids is their head bobbing behavior. Each species has a characteristic sequence of push up or head bob patterns which serve as species recognition mechanisms and defensive signals. Most iguanids have oviparous reproductive habits with eggs usually being laid in late spring or early summer with young hatching in about a month. Many species lay several clutches of eggs in a single summer. Among U.S. iguanids some spiny lizards and some horned lizards are viviparous.

1a Head with bony spines or conspicuous ridge 2

1b Head without bony spines or ridge 8

2a Two large spines at back of head, or head spines small and widely spaced 3

2b Four large head spines at back of head with little or no space between them (Fig. 278) .... Regal Horned Lizard, *Phrynosoma solare* Gray

Figure 278

Figure 278   *Phrynosoma solare*

Dorsum gray with yellow or red tint becoming brownish toward sides with brown blotches along pale light middorsal lines. Basic background color in all *Phrynosoma* species typically varies according to prevailing soil color. Body size 7–10 cm; southern Ariz. southward into Mexico.

2c No head spines, but prominent ridges from snout onto temporal region and across occiput .... Cuban Anole, *Anolis equestris* Merrem

Brown to green with darker blotches on upper dorsum, horizontal yellow stripe extending from shoulder, and yellow blotches on side of head; body size 12–18 cm; introduced at Miami, Fla.

3a Several rows of enlarged pointed scales on each side of the throat between outer chin shields (Fig. 279) .... **Coast Horned Lizard,** *Phyrosoma coronatum* (Blainville)

Figure 279

Figure 279   Chin shields of *P. coronatum*

Yellowish to reddish gray background with dark brown blotches on either side of neck and dark brown wavy spots on each side of middorsal line repeated 3–5 times to base of tail; body size 7–9.5 cm; along Pacific coast in lower two-thirds of Calif. southward in Baja California.

3b Zero to two rows of enlarged scales on each side of throat **4**

4a No enlarged spines on edge of body (Fig. 280) .... **Roundtail Horned Lizard,** *Phynosoma modestum* Girard

Figure 280

Figure 280   *Phrynosoma modestum*

Ground color white, gray to reddish brown with dark blotches on neck and groin which extend forward along sides; body size 3–5 cm. with rare populations having individuals up to 7 cm; western Tex., southern N. Mex. and southeastern Ariz. southward into Mexican Plateau region.

4b Enlarged spines present on edge of body **5**

5a Head spines small, length usually less than diameter of eye .... **Short-Horned Lizard,** *Phrynosoma douglassii* **(Bell)**

Dorsum gray, tan or light brown with pair of dark brown blotches on neck and smaller paired brown blotches on back which may be speckled with white; body size 5–8 cm; widely distributed in higher plateaus and intermountain regions (Fig. 281).

Figure 281

Figure 281   Distribution of *P. douglassii*

5b Head spines long, length 2–4 times diameter of eye **6**

6a Tail flat, dark line down middle of back .... **Flat-tail Horned Lizard,** *Phrynosoma mcalli* **(Hallowell)**

Gray, tan and reddish brown with dark vertebral stripe, light brown blotches and light spots on back; body size 6–8 cm; extreme southeastern Calif. and southwestern Ariz. and adjacent Mexico.

6b Tail not flat, no dark line at middle of back **7**

7a Dark stripes radiating from eye on side of face (Fig. 282).. .... **Texas Horned Lizard,** *Phrynosoma cornutum* **(Hallowell)**

Figure 282

Figure 282   *Phrynosoma cornutum*

Dorsum light brown to gray or reddish, light vertebral stripe, paired dark blotches on neck repeated along back, the posterior margins bordered with white or cream; body size 5–8.5 occasionally up to 10 cm. Distribution (Fig. 283).

Figure 283

Figure 283   Distribution of *P. cornutum*

7b No dark stripes on side of face ....
.... Desert Horned Lizard,
***Phrynosoma platyrhinos* Girard**

Dorsum gray, tan, or brown with paired dark
blotches on neck and wavy dark blotches across
back; body size 6–9 cm. Distribution (Fig. 284).

Figure 284

Figure 284   Distribution of *P. platyrhinos*

8a (1) no ear opening present 9

8b Ear opening present 12

**9a Tail flat with broad black bands on ventral
surface. Two large black bars on side immediately in front of hind legs (Fig. 285)..
.... Greater Earless Lizard,**
***Cophosaurus texanus* Troschel**

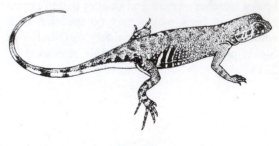

Figure 285

Figure 285   *Cophosaurus texanus*

Gray to light brown with many indistinct light
spots scattered over dorsum, black bars on venter
of tail; body size 5–8.5 cm; central Tex. through
southeastern Ariz. and south into Mexico.

**9b Tail rounded with no black ventral mark-
ings or with small black dots; two small
black bars on sides about midway between
front and hind limbs 10**

10a Small black spots on underside of tail ......
.... Spot-tailed Earless Lizard,
Holbrookia lacerata **Cope**

Figure 286

Figure 286 *Holbrookia lacerata*

Brown with prominent dark brown blotches (Fig. 286) with light borders; female *Holbrookia* have an orange coloration when gravid. Body size 4–7 cm; central and south Tex.

10b No black spots on underside of tail 11

11a Dorsal scales with distinct ridges (keeled) .... Keeled Earless Lizard,
Holbrookia propinqua **Baird & Girard**

Dorsal pattern variable, brown with indistinct dark blotches and light spots, gravid females are orange; body size 4–6 cm; extreme south Tex. into northern Mexico along Gulf coast.

11b Dorsal scales granular or convex, not keeled ....
.... Lesser Earless Lizard,
Holbrookia maculata **Girard**

Color pattern variable, brown with dark blotches and lighter chevrons or dorso-lateral light stripes, profusely covered with small light spots; gravid females are orange; body size 4–7 cm. Distribution (Fig. 287).

Figure 287

Figure 287 Distribution of *H. maculata*

12a Tips of toes expanded 13

12b Tips of toes not expanded 15

13a Dorsum uniformly green or brown sometimes with lighter stripes along midline ....
.. 14

13b Dorsum with darker stripes crossing body in the shape of a reverse chevron, dark stripe between eyes, dewlap (Fig. 288) light yellow..
.... Bahaman Bark Anole,
***Anolis distichus* Cope**

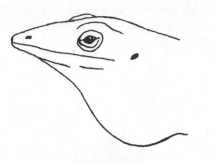

Figure 288

Figure 288   Dewlap of *Anolis*

Color gray to brown with chevron pattern of dark lines on back; body size 3–5 cm; introduced into Miami from Caribbean Islands of Hispaniola and Bahamas.

14a Dewlap pink, dorsal color green or brown
.. Green Anole,
***Anolis carolinensis* Voigt**

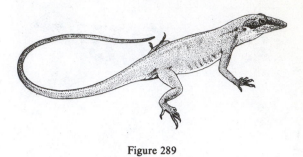

Figure 289

Figure 289   *Anolis carolinensis*

Color changeable from uniform green to brown and combinations in between (Fig. 289); body size 4–7.5 cm. Distribution (Fig. 290).

Figure 290

Figure 290   Distribution of *A. carolinensis*

14b Dewlap red-orange with whitish border
.. Key West Anole,
***Anolis sagrei* Cocteau**

Dorsum brown with occasional lighter spots, females with light stripe along midline of back; body size 3–6 cm; two very similar sub-species have been introduced into Miami and Tampa areas of Fla.

15a Femoral pores absent ....
.... **Northern Curly-tailed Lizard,**
Leiocephalus carinatus **Gray**

Dorsum gray to brown with darker spots on back, tail banded with dark brown particularly toward tip; body size 8–10.5 cm; introduced in isolated locations in southeastern Fla., native of Bahama and Caribbean Islands.

15b Femoral pores present, conspicuous in males **16**

16a A single row of enlarged scales along midline of back (Fig. 291) ....
.... **Desert Iguana,**
Dipsosaurus dorsalis **(Baird & Girard)**

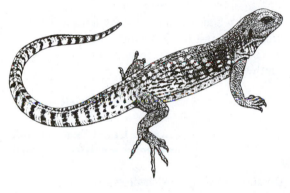

Figure 291

Figure 291   *Dipsosaurus dorsalis*

Dorsum gray to tan with irregular pattern of light brown; body size 10–14 cm; central Ariz. to southern Nev. and southeastern Calif. south into Baja and mainland Mexico.

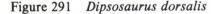

16b No single row of enlarged middorsal scales **17**

17a Transverse gular fold of small scales (Fig. 292a) **18**

17b No transverse gular fold (Fig. 292b) **19**

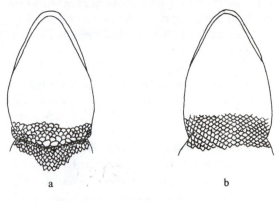

a   b

Figure 292

Figure 292   Gular fold present (a) and absent (b)

18a Conspicuous enlarged fringe of scales on toes **34**

18b No fringe on toes **36**

19a A pocket in the skin just behind rear legs, males have pink belly patches ....
.... **Rosebelly Lizard,**
Sceloporus variabilis **Wiegmann**

Dorsum tan to olive-brown with row of brown blotches down each side of back bordered by continuous dorsolateral light stripe; body size 4–5.5 cm; south Tex. extending south into Mexico.

19b No pocket in the skin behind rear legs, males without or usually with blue belly patches (belly patches absent in *S. virgatus*) **20**

20a Lateral scales small and granular ....
.... Canyon Lizard,
Sceloporus merriami **Stejneger**

Dorsum gray to light brown, vertical black bar on shoulder, rows of dark spots on back with variable light spotting; body size 4–5.5 cm; Stockton Plateau and lower Big Bend of western Tex. and adjacent Mexico.

20b Lateral scales overlapping, not granular
.. 21

21a Supraoculars very large, in a single row with posterior ones not separated from median head scales by row of smaller scales
.. 22

21b Supraoculars smaller, in single or double row always separated from median head scales by at least one complete row of small scales 25

22a No black shoulder patch, body dark blue to black on dorsum and venter ....
.... Granite Spiny Lizard,
Sceloporus orcutti **Stejneger**

Dorsum uniform dark or with indistinct darker crossbands and wedge-shaped mark on either side of neck; body size 8–10 cm; extreme southern Calif. south through Baja California.

22b Distinct black shoulder patch 23

23a Forelegs with crossbars of brown (Fig. 293)
.... Clark's Spiny Lizard,
Sceloporus clarki **Baird & Girard**

Figure 293

Figure 293   *Sceloporus clarki*

Gray with bluish-green tint, with irregular darker crossbands and black shoulder patch; body size 9–13 cm; central Ariz. to southwestern N. Mex. southward into Mexico.

23b No crossbars of brown on forelegs, body color not tinted green ....
.... Desert Spiny Lizard,
Sceloporus magister **Hallowell**

Dorsum tan to yellowish-brown with crossbands or blotches of darker brown, black shoulder patch; body size 9–14 cm. Distribution (Fig. 294).

Figure 294

Figure 294   Distribution of *Sceloporus magister*

24a Scales on back of thigh granular, noticeably smaller than scales on front of thigh 25

24b Scales on back of thigh not granular nor smaller than scales on front of thigh ... 27

25a Scales on side of neck contrastingly smaller than scales on top of neck .... Mesquite Lizard, *Sceloporus grammicus* Wiegmann

Gray to olive with irregular dark wavy lines across dorsum (more prominent in females); body size 5–7.5 cm; enters U.S. only in extreme southern tip of Tex.

25b Scales on side of neck not contrastingly smaller than scales on top of neck 26

26a Femoral pores separated by few scales (1–3) .... Bunch Grass Lizard, *Sceloporus scalaris* Wiegmann

Figure 295

Figure 295 *Sceloporus scalaris*

Tan or light brown, patternless or with light dorsolateral light stripe often intersecting with rows of dark brown chevron marks (Fig. 295); body size 3–5.5 cm; mountains of southeastern Ariz. and adjacent N. Mex. southward into Mexico.

26b Femoral pores separated by more than 3 scales .... Sagebrush Lizard, *Sceloporus graciosus* Baird & Girard

Dorsum gray or brown with darker blotches or irregular crossbars and dorsolateral light stripes; body size 4–6.5 cm. Distribution (Fig. 296).

Figure 296

Figure 296 Distribution of *S. graciosus*

27a Black collar around neck 28

27b No black collar around neck 30

28a Tail banded with contrasting dark and light bands, supraoculars in two longitudinal rows 29

28b Tail not banded, supraoculars in one row .... Yarrow's Spiny Lizard, *Sceloporus jarrovi* Cope

Figure 297

Figure 297 *Sceloporus jarrovi*

Dorsum dark brown, charcoal or black with irregular darker blotches in females or lighter centers of scales in males, black neck band bordered with white (Fig. 297); body size 6–9.5 cm; mountains of southeastern Ariz. and southwestern N. Mex. southward into Mexico.

29a Tail bands more distinct toward tip (Fig. 298), dorsal color tinted tan, gray or red .... Crevice Spiny Lizard, *Sceloporus poinsetti* Baird & Girard

Figure 298

Figure 298 *Sceloporus poinsetti*

Dorsum tan, gray or reddish brown with dark blotches or bands becoming more distinct posteriorly grading into black tail-bands, broad black collar bordered with white; body size 7–13 cm; central Tex. to southwestern N. Mex. southward onto Mexican Plateau.

29b Tail bands less distinct toward tip, dorsal color tinted blue .... Blue Spiny Lizard, *Sceloporus cyanogenys* Cope

Brown to gray with greenish blue tint particularly in males, black collar bordered with white; white spots on back; body size 8–15 cm; extreme southern Tex. southward into Mexico.

30a Posterior surface of thigh light with very small dark spots, dorsal scales larger .... Texas Spiny Lizard, *Sceloporus olivaceus* Smith

Gray to light brown, wavy dark irregular bands across back, faint dorsolateral light stripe (more prominent in males); body size 6–12 cm; north central Tex. southward into Mexico.

30b Posterior surface of thigh with longitudinal dark line; dorsal scales smaller 31

31a Distinct lateral stripe (dark brown or white) 32

31b Lateral stripe (if present) white with indistinct borders 33

32a Lateral stripe white or cream color ....
.... Striped Plateau Lizard,
Sceloporus virgatus Smith

Rusty brown with flecks of white and dark brown, dorsolateral light stripe and lateral light stripe with well defined borders; *males without blue belly patches;* body size 4–6 cm; mountains of extreme southeastern Ariz. and southwestern N. Mex. southward into Mexico. Some populations of the prairie fence lizard *(Sceloporus undulatus garmani)* have striping like *S. virgatus* but males have blue belly patches.

32b Lateral stripe dark brown ....
.... Florida Scrub Lizard,
Sceloporus woodi Stejneger

Brown or grayish with broad dark brown lateral stripe, females with irregular dark brown markings on back; body size 4–5.5 cm; isolated sandy areas on Fla. peninsula.

33a Scales on posterior surface of thigh abruptly smaller than on dorsal surface of thigh, males with single large blue throat patch .... Western Fence Lizard,
Sceloporus occidentalis Baird & Girard

Charcoal or brown back with irregular dark marks, males with blue or greenish tint; body size 7–9 cm. Distribution (Fig. 299).

Figure 299

Figure 299 Distribution of *Sceloporus occidentalis*

33b Scales on posterior surface of thigh becoming gradually smaller than those on dorsal surface of thigh, males with blue patches on each side of throat .... Eastern Fence Lizard, *Sceloporus undulatus* (Bosc in Latreille)

Body gray or brown, pattern variable from wavy dark bands or with dorsolateral light stripes usually not as distinct as in *S. virgatus;* body size 4–7.5 cm. Distribution (Fig. 300).

Figure 300

Figure 300   Distribution of *S. undulatus*

34a (18) Prominant black spots or bars on sides of belly 35

34b Black spots on belly very small or absent .... Coachella Valley Fringe-toed Lizard, *Uma inornata* Cope

Black reticulate pattern with white to gray donut shaped marks, black net forming streaks on neck and shoulders; body size 8–12 cm; restricted to sand dunes in Coachella Valley in southern Calif.

35a Dark marks on throat meet at middle (at least posteriorly) forming a U or V........... .... Mohave Fringe-toed Lizard, *Uma scoparia* Cope

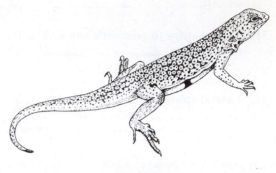

Figure 301

Figure 301   *Uma scoparia*

Black reticulum (Fig. 301) with white to gray donut shaped marks, black marks becoming dots on neck and shoulders; body size 7–10 cm; restricted to sand dunes in extreme southern Calif.

35b Marks on throat do not meet at middle, but form streaks .... **Colorado Desert Fringe-toed Lizard,** *Uma notata* **Baird**

Black and white to gray dorsum, netlike with donut shaped marks, black net forming streaks on shoulders and neck; body size 7–10 cm; sand dunes in extreme southern Calif., southwestern Ariz., and northeastern Baja (Fig. 302).

Figure 302

Figure 302　Distribution of *Uma notata*

36a (18) Tail flattened with black bands particularly conspicuous on venter .... **Zebratail Lizard,** *Callisaurus draconoides* **Blainville**

Figure 303

Figure 303　*Callisaurus draconoides*

Gray to brown above with numerous light spots and irregular dark marks (Fig. 303). Two black bars on side and black bars on venter of tail; body size 6–9 cm; southwestern U.S. from extreme western N. Mex. to north central Nev. and south into Baja California and Sonora, Mexico.

36b Tail not flattened and without black bands **37**

37a Superciliary scales not overlapping, rostral subdivided into small scales...................... .... Chuckwalla, *Sauromalus obesus* Baird

39a Enlarged supraoculars in 2 or 3 rows .... Banded Rock Lizard, *Petrosaurus mearnsi* (Stejneger)

Figure 304

Figure 304   *Sauromalus obesus*

Gray to olive with black crossbands in juveniles and females (Fig. 304); males often with reddish brown blotches; body size 13–21 cm; deserts of southeastern Calif., western Ariz., southern Nev. and southwestern Utah.

37b Superciliaries overlapping and a single scale (rostral) at tip of upper lip 38

38a Head scales small and granular except occasionally over eyes 43

38b Head scales consisting of enlarged plates 39

Figure 305

Figure 305   *Petrosaurus mearnsi*

Dorsum gray to olive with light spots, single black collar and irregular dark crossbands and blotches on body continuing on tail where they form prominant bands (Fig. 305); body size 8–10 cm; extreme southern Calif. and southward into Baja California.

39b Enlarged supraoculars in single row 40

40a Supranasals present, conspicuous dark blue axillary spot .... Side-blotched Lizard, *Uta stansburiana* Baird & Girard

Color pattern highly variable, tan with brown blotches or brown with light spots on dorsolateral light stripes (or chevrons), males with blue flecks, conspicuous axillary dark blotch; body size 3.5–6 cm. Distribution (Fig. 306).

Figure 306

Figure 306   Distribution of *Uta stansburiana*

40b Supranasals absent, no axillary spot .. 41

41a Dorsal scales uniformly small in size .... Small-scaled Lizard, *Urosaurus microscutatus* (Van Denburgh)

Dorsum gray with dark blotches and scattered light spots; body size 3–5 cm; extreme southern Calif. southward into Baja California.

41b Dorsal scales not uniform in size, series of enlarged scales on top of back 42

42a Broad band of enlarged scales along middle of back (Fig. 307a) .... Brush Lizard, *Urosaurus graciosus* Hallowell

a

b

Figure 307

Figure 307   Middorsal scales of *Urosaurus graciosus* (a) and *Urosauras ornatus* (b)

Uniform gray to brownish dorsum or with fine transverse dark bars; body size 4–6 cm; southwestern Ariz., southeastern Calif., and extreme southern Nev., south to northern end of Gulf of California in Sonora and Baja California.

**42b Enlarged scales separated into two longi-
tudinal series by series of small scales at
midline (Fig. 307b)..............** Tree Lizard,
Urosaurus ornatus **(Baird & Girard)**

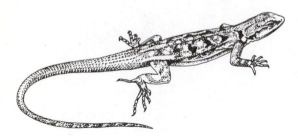

Figure 308

Figure 308   *Urosaurus ornatus*

Dorsum gray or brown with dark blotches, longitudinal lines or crossbars (Fig. 308); body size
4–6 cm. Distribution (Fig. 309).

Figure 309

Figure 309   Distribution of *U. ornatus*

43a (38) One or two black bands across neck
**.. 46**

43b No black bands across neck................. 44

**44a Large dark spots in rows on back ....
.... Reticulate Collared Lizard,**
Crotaphytus reticulatus **Baird**

Grayish or reddish brown with rows of darker
spots, light reticulate pattern on dorsum, males
with dark bars on side of head; body size 10–14
cm; extreme southern Tex. and adjacent Mexico.

**44b Dark spots (leoparine pattern) irregular in
size and distribution, not in rows 45**

**45a Gray longitudinal streaks on throat; head
length (tip of snout to anterior border of
ear) greater than head width ....
.... Longnose Leopard Lizard,**
Gambelia wislizenii **(Baird & Girard)**

Figure 310

Figure 310   *Gambelia wislizenii*

Brown spots (Fig. 310) on gray, brown or reddish
ground color, light lines sometimes reticulate or
forming bars; body size 9–12 cm. Some authors

place this species in the genus *Crotaphytus*. Distribution (Fig. 311).

Figure 311

Figure 311 Distribution of *G. wislizenii*

45b Gray blotches on throat; head length equal or less than head width ....
.... Bluntnose Leopard Lizard,
Gambelia silus (Stejneger)

Similar to long-nosed leopard lizard but often redder and large dark blotches under spots more prominent; body size 9–12 cm; San Joaquin Valley in Calif. Some authors place this species in *Crotaphytus*.

46a Two black bands across neck, first often incomplete dorsally .... Collared Lizard,
Crotaphytus collaris (Say)

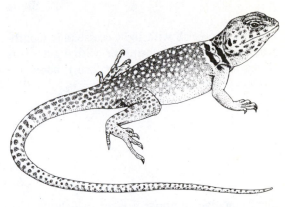

Figure 312a

Figure 312a *Crotaphytus collaris*

Color variable, greenish, brownish, or yellowish with many light spots (Fig. 312a); body size 9–11 cm. Distribution (Fig. 312b).

Figure 312b

Figure 312b Distribution of *C. collaris*

46b **Black bands across neck usually reduced; both incomplete dorsally** ....
.... **Black-collared Lizard,**
Crotaphytus insularis **Van Denburgh & Slevin**

Figure 313

Dorsum brownish with light crossbands (southern California-Baja form) or with light spots (Great Basin-Colorado Desert form); body size 8–10 cm. Distribution (Fig. 313).

Figure 313   Distribution of *C. insularis*

SKINK FAMILY
Scincidae

The Scincidae comprise a large, widely distributed family most abundant and diverse in the Australian and Oriental regions of the world. Fifteen species are found in the United States. Most skinks are terrestrial and typically inhabit leaf litter and humid microhabitats of forests although some (e.g., Broad-head Skink) are arboreal. Skinks have smooth shiny scales which are rather uniform in appearance and size. The tongue is extensible and the legs are often small in relation to the size of the body. The tail is fragile and easily broken but is rapidly regenerated if lost. Some skinks have been shown to lose the tail to a predator but return to the site of the attack and eat the lost tail thus recovering the energy which is stored in the often thick tail. Few other species of lizards are known to display such behavior. The brightly colored tail of many juvenile skinks (often blue or red) has been shown to be important in age discrimination but may also be important in predator distraction. Skinks are typically diurnal although secretive in their habits. They are insectivorous although some larger species may eat other small lizards and snails. Some skinks are viviparous but all U.S. species are oviparous. A few species exhibit parental care of brooding the eggs with the female maintaining constant guard over the clutch of eggs until the last egg has hatched. The heads of males in many species become tinted with red or orange during the breeding season.

1a **Legs reduced with not more than two toes** .... **Sand Skink,**
Neoseps reynoldsi **Stejneger**

Gray to yellowish brown; body size 4–6.5 cm; restricted to isolated sandy localities in central Fla.; this is a burrowing species with wedge-shaped head.

1b **Legs may be reduced but each always with five digits** **2**

2a No supranasals, lower eyelid with trans-
lucent window .... Ground Skink,
Scincella laterale (Say)

Figure 314

Figure 314   *Scincella laterale*

Dorsum gold to gray-brown with dark brown
dorsolateral stripe (Fig. 314); body size 3–5 cm;
there has been considerable debate about the
proper genus for this species and various references may list it in the genera *Lygosoma* or
Leiolopisma. Distribution (Fig. 315).

Figure 315

Figure 315   Distribution of *S. laterale*

2b Supranasals present 3

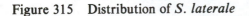

3a Scale rows on side of body are oblique to
those on back (Fig. 316a) ....
.... Great Plains Skink,
Eumeces obsoletus (Baird and Girard)

Figure 316a

Figure 316a   Oblique scale rows of *Eumeces
obsoletus*

Figure 316b

Figure 316b   Parallel scale rows of *Eumeces*
species

Figure 317

Figure 317   *Eumeces obsoletus*

Ground color of light tan, cream or gray with each
scale having a black or brown border (Fig. 317),
young are entirely black with blue tails; body size

8–14 cm; southcentral plains states from western and southern Nebr. to southern Tex., westward to central Ariz.

3b Scale rows on side of body are parallel to those on back (Fig. 316b) 4

**4a Light Y-shaped mark on top of head ending on neck (Fig. 318)..............................
.... Mountain Skink,
Eumeces callicephalus Bocourt**

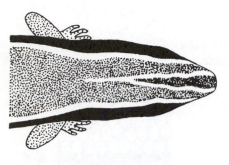

Figure 318

Figure 318 *Eumeces callicephalus*

Olive tan to brown above with pale light line extending posteriorly from eye with dark brown below; body size 5–6.5 cm; restricted to mountains of extreme southeastern Ariz. and southward into Mexico.

4b No light Y-shaped mark on top of head or, if present, it is continuous with middorsal light stripe 5

5a Dorsolateral light stripes if present occur on the second scale row from middorsal line 6

5b Dorsolateral light stripes not present or if present do not include the second scale row 8

**6a No postnasal scales .... Mole Skink,
Eumeces egregius (Baird)**

Color gray or brown with longitudinal light stripes, tail red, yellow, orange, blue, or brown; body size 3–6 cm; southeastern Ala. and southern Ga. southward to Fla. Keys.

6b Postnasal scale(s) present 7

**7a Adults not distinctly striped, young with red (or blue) tail and dark lateral line stopping at base of tail .... Gilbert's Skink,
Eumeces gilberti Van Denburgh**

Olive or brown with light stripes in young which are gradually lost until adults with almost uniform ground color and variable dark spotting which tends to form lines, tail becomes red with age; body size 5–12 cm; lower two-thirds of central Calif., southern Nev. and central Ariz.

**7b Adults distinctly striped, young with blue tail; dark lateral line extending well beyond base of tail (Fig. 319) ....
.... Western Skink,
Eumeces skiltonianus (Baird & Girard)**

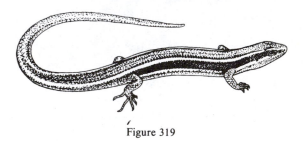

Figure 319

Figure 319 *Eumeces skiltonianus*

Brown middorsal band edged with black and bordered with dorsolateral light stripes, blue tail fades with age; body size 5–8 cm. Distribution (Fig. 320).

Figure 320

Figure 320   Distribution of *E. skiltonianus*

8a No dorsolateral light stripes ....
Eumeces **sp.**

Parts of several species may key to here. Adult *E. gilberti* lose the striped pattern as do adult males of *E. fasciatus, E. inexpectatus,* and *E. laticeps;* also, there is a non-patterned morph of *E. multivirgatus.* Identification can generally be made with ease using a combination of locality and body size, see species descriptions elsewhere in key.

8b Pattern consisting of variable numbers of stripes **9**

9a Dorsolateral light lines extending full length of body on third scale rows from middorsal line .... **Many-lined Skink,** *Eumeces multivirgatus* **(Hallowell)**

Dorsal pattern variable (unicolored morph in some populations) but usually alternating dark and light longitudinal stripes; body size 4–6 cm. Distribution (Fig. 321).

Figure 321

Figure 321   Distribution of *E. multivirgatus*

9b Dorsolateral stripes not restricted to third scale rows from middorsal line (usually involving fourth or fourth and fifth rows only) **10**

10a Postnasal scale(s) present (Fig. 322a) .. 11

10b No postnasals (Fig. 322b) 13

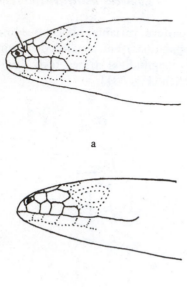

Figure 322

Figure 322   Postnasal scale present (a) and absent (b)

11a Middle row of scales under base of tail are enlarged (Fig. 323a) 12

11b Middle row of scales under base of tail not enlarged (Fig. 323b) ....
.... Southeastern Five-lined Skink,
***Eumeces inexpectatus* Taylor**

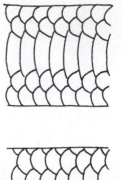

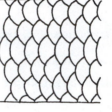

Figure 323

Figure 323   Subcaudal scales enlarged (a) and not enlarged (b)

Dark brown to black dorsum with five longitudinal light stripes which tend to fade with age (particularly in males) as does the blue tail; body size 4–9 cm. Distribution (Fig. 324).

Figure 324

Figure 324   Distribution of *E. inexpectatus*

12a Four anterior upper labials and two post-labials (Fig. 325a)..............................
.... Five-lined Skink,
Eumeces fasciatus (Linnaeus)

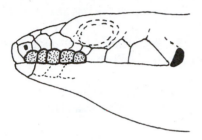

Figure 325a

Figure 325a *Eumeces fasciatus*

Figure 325b

Figure 325b *Eumeces laticeps*

Five longitudinal light stripes on black background, young with blue tails which fade with age as do light stripes; body size 4–9 cm. Distribution (Fig. 326).

Figure 326

Figure 326 Distribution of *E. fasciatus*

12b Five anterior upper labials and no postlabials (Fig. 325b)..............................
.... Broadhead Skink,
Eumeces laticeps (Schneider)

Five light lines on dark brown to olive background, blue tail in young fades with age, as do the light stripes; large adult males develop red head; body size 7–14 cm. Distribution (Fig. 327).

Figure 327

Figure 327 Distribution of *E. laticeps*

13a Dorsolateral light lines extending well onto tail 15

13b Dorsolateral light lines not extending beyond groin 14

14a Dorsolateral light lines extend to groin .... **Four-lined Skink,** *Eumeces tetragrammus* **(Baird)**

Dorsum brown to gray with 4 longitudinal light stripes tending to fade with age, tail blue in young; body size 4–7.5 cm; extreme southern Tex. to Veracruz, Mexico.

14b Dorsolateral light lines ending before midbody .... **Short-lined Skink,** *Eumeces brevilineatus* **Cope**

Figure 328

Figure 328   *Eumeces brevilineatus*

Four longitudinal light stripes extending just beyond shoulder (Fig. 328) on background of brown, gray or olive, blue tail in young; body size 4–7 cm; central and western Tex. (and adjacent Mexico), and southern Ariz.

15a One postmental scale (Fig. 329a), dorsolateral light line on 3rd and 4th scales from midline .... **Coal Skink,** *Eumeces anthracinus* **(Baird)**

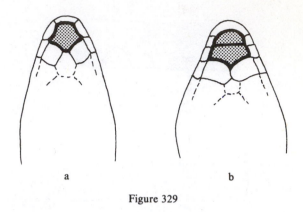

a   b

Figure 329

Figure 329   Postmentals in *E. anthracinus* (a) and *E. septentrionalis* (b)

Gray to brown with four longitudinal light stripes, tail purplish; body size 4–7 cm. Distribution (Fig. 330).

Figure 330

Figure 330   Distribution of *E. anthracinus*

15b Two postmental scales (Fig. 329b), dorso-lateral light line on 5th and/or 4th scale row from midline .... Prairie Skink, *Eumeces septentrionalis* (Baird)

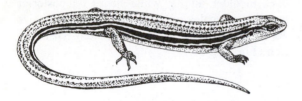

Figure 331

Figure 331   *Eumeces septentrionalis*

Four or seven longitudinal light stripes on brown or olive background (Fig. 331), blue tail in young fades with age; body size 5–8 cm. Distribution (Fig. 332).

Figure 332

Figure 332   Distribution of *E. septentrionalis*

ALLIGATOR LIZARD FAMILY
Anguidae

The anguids of the United States are often called the lateral fold lizards due to the unique characteristic of a fold of skin along the side of the body which is covered with small granular scales greatly different from the larger dorsal and ventral plates. Oddly enough, only the two U.S. genera possess the lateral fold while other members of the family do not. The anguids are widely distributed particularly in the tropics but comprise a fairly small number of species. The anguid body form may be typically lizard-like or may be snake-like (legless) as in the glass lizard of the genus *Ophisaurus* of the eastern U.S. The glass lizards may be easily distinguished from snakes by the presence of eyelids and ear openings. The tongue is extensible and bifurcate. Most species are secretive in their habits and are seldom seen even by experienced observers. As far as known most are insectivorous. Some species are oviparous while others are viviparous.

1a With four legs **4**

1b With no legs **2**

2a Upper labials contact orbit of eye at least for 1 or 2 scales (Fig. 333a) .... Island Glass Lizard, *Ophisaurus compressus* Cope

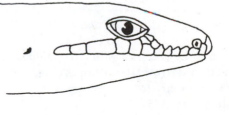

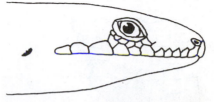

Figure 333

Figure 333   *Ophisaurus compressus* (a) and other *Ophisaurus* (b)

Tan with dark stripe on each side at scales 3 &
4 above lateral groove, middorsal dark stripe occasionally broken into a series of dashes; body size
10–17 cm, total length to 61 cm; Fla. (except
southern tip) and southeastern coasts of Ga. and
S.C.

**2b Upper labials do not contact orbit of eye
(Fig. 333b)** **3**

**3a Distinct middorsal dark stripe or dashes
present** .... Slender Glass Lizard,
Ophisaurus attenuatus Baird

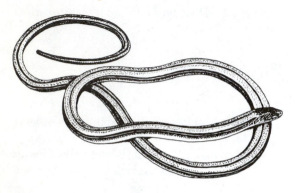

Figure 334

Figure 334   *Ophisaurus attenuatus*

Golden tan to light brown with dark middorsal
stripe and dark narrow lateral stripes (Fig. 334);
body size 20–29 cm, total length up to 105 cm.
Distribution (Fig. 335).

Figure 335

Figure 335   Distribution of *O. attenuatus*

3b No distinct middorsal dark stripe ....
.... **Eastern Glass Lizard,**
Ophisaurus ventralis **(Linnaeus)**

Dorsum tan to green with small longitudinal dark
lines in older adults, white flecks on posterior
margins of scales on back and neck; body size
20–30 cm, total length up to 110 cm. Distribution (Fig. 336).

Figure 336

Figure 336   Distribution of *O. ventralis*

4a A small median scale immediately behind rostral (Fig. 337a).............................
.... **Texas Alligator Lizard,**
Gerrhonotus liocephalus **Weigmann**

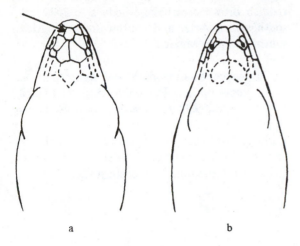

Figure 337

Figure 337 Median postrostral scale in *G. liocephalus* compared to double postrostrals in *Elgaria*

Light brown to reddish brown dorsum with irregular bars of white or dark brown scales across body (Fig. 338); body size 15–20 cm; central Tex. westward to Big Bend and southward into Mexico.

Figure 338

Figure 338 *Gerrhonotus liocephalus*

4b No median scale behind rostral (Fig. 337b)
.... **5**

5a No distinct crossbands on body ....
.... **Northern Alligator Lizard,**
Elgaria coeruleus **(Weigmann)**

Olive to bluish brown with darker irregular small spots sometimes forming rows on sides; body size 9–13 cm. Some authors consider *Elgaria* and some Middle American anguids to be *Gerrhonotus*. Distribution (Fig. 339).

Figure 339

Figure 339 Distribution of *Elgaria coeruleus*

5b Distinct crossbands on body **6**

6a Dark crossbands on body as wide or wider than lighter ground color **7**

6b Dark crossbands narrower than lighter ground color ....
.... **Southern Alligator Lizard,**
Elgaria multicarinatus **(Blainville)**

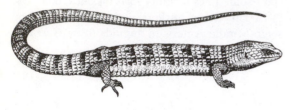

Figure 340

Figure 340 *Elgaria multicarinatus*

Reddish brown to gray dorsum with dark brown narrow crossbands (Fig. 340), white flecks on sides; body size 12–14 cm; western coast from northern Baja California to northern Calif.

7a Contrasting dark and light spots on upper lip .... **Arizona Alligator Lizard,**
Elgaria kingi **Gray**

Light brown to gray with broad dark brown or reddish brown crossbands; body size 7–12 cm; southeastern Ariz. and southwestern N. Mex. southward into Mexico.

7b No contrasting dark and light spots on upper lip ...... **Panamint Alligator Lizard,**
Elgaria panamintinus **(Stebbins)**

Light yellow to tan with broad brown crossbands (much darker in young); body size 9–15 cm; restricted to Panamint Mts. eastern Calif.

LEGLESS LIZARD FAMILY
Anniellidae

This family contains only two species of the single genus *Anniella* which is distributed in coastal and central Calif. from near San Francisco Bay to northern Baja California. The body of *Anniella* is elongate, about the size of a pencil, and has no appendages although vestiges of the pelvic girdle and clavicle are present. The body is uniformly covered by smooth, overlapping scales with rounded posterior borders. The tail is blunt and stout and is used as a pick to move the body backwards. The lizards live in sandy soil, leaf litter, or under rocks and logs, and seldom venture above ground. Fossorial adaptations include the limbless body, stout tail, shovel shaped head, and countersunk jaws. Feeding primarily on soft bodied and small insects or larvae, these lizards give birth to 1–4 young in late summer.

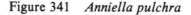

Figure 341

Figure 341 *Anniella pulchra*

California Legless Lizard, *Anniella pulchra* Gray, yellowish brown to olive brown above (one subspecies dark brown to black in Monterey Peninsula, Calif.) with dark middorsal line and 2–3 dark lines on side (Fig. 341); body size 10–15 cm.

NIGHT LIZARD FAMILY
Xantusiidae

The night lizards comprise a small family of new world lizards occurring from the southwestern United States to Panama and Cuba. Two of the five genera occur in the U.S. including three species. Eyelids are absent and the skin is soft with small granular scales on the back and large quadrangular scales on the belly. The head is covered with large flat scales. These lizards are secretive and nocturnal and largely insectivorous although other small arthropods and some plant materials are eaten. The tail is easily broken and regenerates rapidly. The night lizards are long-lived and viviparous giving birth to living young in late summer or early fall.

1a Two rows of supraocular scales (Fig. 342a) and 16 rows of ventral scales across belly .... **Island Night Lizard,** *Klauberina riversiana* **(Cope)**

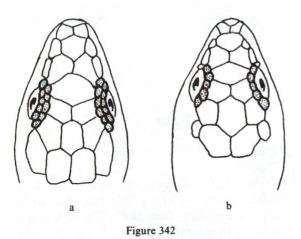

a   b

Figure 342

Figure 342   Supraoculars in two rows (a) in *Klauberina* compared to one row (b) in *Xantusia*

Color variable but generally consisting of a dark brown or black reticulation over a gray to reddish brown ground color; body size 7–10 cm; occurs only on islands off the coast of Calif. (San Clemente, Santa Barbara, San Nicholas).

1b One row of supraocular scales (Fig. 342b) **2**

2a Fourteen rows of ventral scales across belly .... **Granite Night Lizard,** *Xantusia henshawi* **Stejneger**

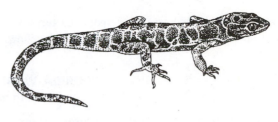

Figure 343

Figure 343   *Xantusia henshawi*

Color consisting of large brown spots on tan ground color (Fig. 343); body size 4–7 cm; extreme southern Calif. and northern Baja California.

2b Twelve rows of ventral scales across belly .... **Desert Night Lizard,** *Xantusia vigilis* **Baird**

(Fig. 344); body size 3–5 cm. Distribution (Fig. 345).

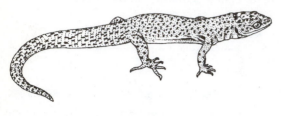

Figure 344

Figure 344   *Xantusia vigilis*

Small dark brown irregular spots scattered on lighter ground color of light tan, gray or yellow

Figure 345

Figure 345   Distribution of *X. vigilis*

FLORIDA WORM LIZARD FAMILY[1]
Rhineuridae

The Florida Worm Lizard, *Rhineura floridana,* is the only living example of the family Rhineuridae. This and other worm lizard families are grouped into the Amphisbaenia and considered with the lizards although most authors now consider them as equivalent to the lizard (Lacertilia) and snake (Serpentes) groups (see page 119). Worm lizards bear remarkable resemblance to earth worms because of the rings of flat scales which surround the limbless body (a few non-U.S. species possess anterior limbs). The amphisbaenia are distributed in tropical and subtropical regions in the western hemisphere and in Africa and the Mediterranean region. All species are burrowers with short usually flat tails, no ear openings and eyes concealed under the skin of a stout spade-like head with recessed jaws. Food consists of worms and soft bodied insects. Species may bear living young or lay eggs.

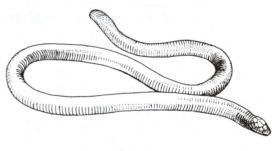

Figure 346

Figure 346   *Rhineura floridana*

Florida Worm Lizard, *Rhineura floridana* (Baird) color is uniformly light rose (Fig. 346); body size 20–28 cm; from Gulf to Atlantic coasts across north central Fla.

[1]Also known as Graveyard Lizards.

BEADED LIZARD FAMILY
Helodermatidae

The beaded lizards are the only venomous lizards and include the Gila Monster of the southwestern deserts of the United States and the Mexican Beaded Lizard of western Mexico to Guatemala. The poison glands, representing modified sublingual salivary glands, are located in the lower jaw rather than in the upper jaw as in snakes. Unlike snakes, the glands are not directly connected to the teeth but release the poison from ducts along the tooth ridge. Grooves on the teeth help draw the poison into the wound. The bite is painful but rarely fatal to man. Food consists of eggs of birds and reptiles, small mammals and occasionally other lizards. They are active in late evening and night and are egg layers.

lations alternating with patches of yellow-orange or red (Fig. 347); body size 30–40 cm. Distribution (Fig. 348).

Figure 348

Figure 348   Distribution of *H. suspectum*

Figure 347

Figure 347   *Heloderma suspectum*

Gila Monster, *Heloderma suspectum* Cope, color consisting of irregular black patches or reticu-

WHIPTAIL FAMILY
Teiidae

The whiptails of the United States represent a single genus of a large family of New World lizards whose greatest abundance and diversity occurs in South America. The whiptails are characterized by a long slender body, elongate tail and pointed head. They have extensible forked tongues. The body is covered with small granular scales above and large quadrangular plates arranged in rows on the venter. The head is covered with large shields. Most forms are extremely rapid runners. Whiptails are insectivorous and

typically find insects by turning sticks and stones and scratching the ground while actively foraging for food. All U.S. species are oviparous and most occur in the desert grasslands of the southwest. The U.S. genus *Cnemidophorus* contains several parthenogenetic species of which some are difficult to distinguish. Species of whiptails continue to be described and the student should not anticipate the following account to remain adequate.

1a Dorsal color pattern consisting of distinct longitudinal light stripes to groin; may have light spots between lines but lines not disrupted, interconnected, nor with transverse bars 6

1b Dorsal color pattern variable but not consisting of distinct longitudinal lines to groin or if lines are apparent they are interconnected with cross-lines or have longitudinal bars 2

2a Mesoptychial (those in front of gular fold) scales small and gradually grade into granular scales of fold .... Western Whiptail, *Cnemidophorus tigris* Baird & Girard

Figure 349

Figure 349   *Cnemidophorus tigris*

Brown or tan with black spots or crossbands (Fig. 349); body size 7–9 cm. Distribution (Fig. 350).

Figure 350

Figure 350   Distribution of *C. tigris*

2b Mesoptychial scales noticeably enlarged abruptly at gular fold 3

3a Circumorbital scale row extends far forward separating third supraocular from frontal and frontoparietal scales (Fig. 351a) 4

3b Circumorbital scale row not extending far forward, not separating third supraocular from frontal and frontoparietal scales (Fig. 351b) 5

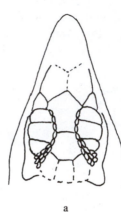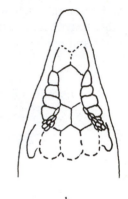

a   b

Figure 351

Figure 351   Circumorbitals extending (a) or not extending (b) far forward

4a Dorsal granules around midbody (DGAB) 94–108; dorsal color pattern consisting of squares or longitudinal lines completely disrupted by horizontal bars ....
.... Gray-checkered Whiptail, *Cnemidophorus dixoni* Scudday

Network of white lines and crossbars over black; rusty rump, *all female species;* body size 7–10 cm; isolated populations in Texas Big Bend and southwestern N. Mex.

4b DGAB 77–98; dorsal color pattern not consisting of squares, longitudinal lines with lateral projections but not completely disrupted .... **Colorado Checkered Whiptail,** *Cnemidophorus tesselatus* (Say)

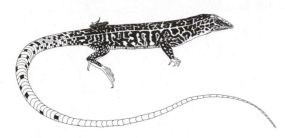

Figure 352

Figure 352 *Cnemidophorus tesselatus*

Light lines with bars and spots on black (Fig. 352), rusty rump; body size 7–10 cm. Distribution (Fig. 353).

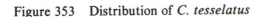

Figure 353

Figure 353 Distribution of *C. tesselatus*

5a Dorsal pattern of light spots (Fig. 354a), rusty red color on back generally confined to anterior three fourths of dorsum .... **Canyon Spotted Whiptail,** *Cnemidophorus burti* Taylor

a   b

Figure 354

Figure 354 *C. burti* (a); *C. septemvittatus* (b)

Light spots on darker background with reddish head and neck, light stripes in young; body size 8–14 cm; southeastern Ariz.

5b Dorsal pattern of light lines with transverse bars and spots (Fig. 354b); rusty color confined to posterior quarter of dorsum.... .... **Plateau Spotted Whiptail,** *Cnemidophorus septemvittatus* Cope

Light lines with bars and spots, rusty rump; body size 8–10.5 cm; Texas Big Bend south into Mexico.

6a (1) No light spots in dark dorsum field between light longitudinal lines.................. 7

6b Light spots in dark field at least between some of the light longitudinal lines 11

7a One frontoparietal scale (Fig. 355a) ....
.... **Orangethroat Whiptail,**
Cnemidophorus hyperythrus **Cope**

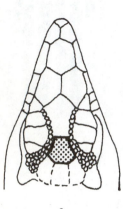 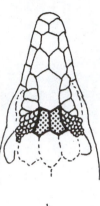

a   b

Figure 355

Figure 355   Frontoparietal, one scale (a) or two scales (b)

Light stripes on dark brown; orange throat; body size 5–7 cm; southwest Calif. south into Baja California.

7b Two frontoparietal scales (Fig. 355b) **8**

8a Mesoptychial scales granular or only slightly enlarged (Fig. 356a) ....
.... **Little Striped Whiptail,**
Cnemidophorus inornatus **Baird**

Six to seven light stripes on brown, no spots, bluish tail; body size 5–7.5 cm; N. Mex., western Tex. south into Mexico.

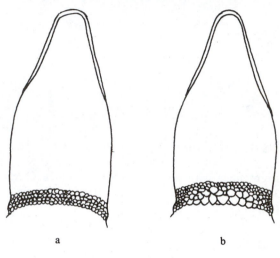

a   b

Figure 356

Figure 356   Mesoptychial scales small (a) or enlarged (b)

8b Mesoptychial scales noticeably enlarged (Fig. 356b) **9**

9a Postantebrachial scales (scales of forearm) granular or only slightly enlarged .... Six-lined Racerunner, *Cnemidophorus sexlineatus* (Linnaeus)

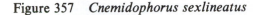

Figure 357

Figure 357 *Cnemidophorus sexlineatus*

Six to seven light stripes on black (Fig. 357), prairie race bright green anterior; body size 5–8 cm. Distribution (Fig. 358).

Figure 358

Figure 358 Distribution of *C. sexlineatus*

9b Postantebrachial scales enlarged **10**

10a Enlarged preanal scales 3 or less (Fig. 359), tail olive or bluish green .... Desert Grassland Whiptail, *Cnemidophorus uniparens* Wright & Lowe

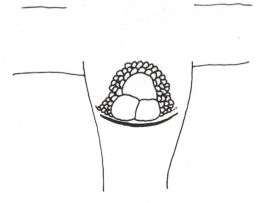
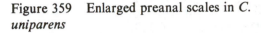

Figure 359

Figure 359 Enlarged preanal scales in *C. uniparens*

Six or seven light stripes on dark brown, tail olive to bluish green, *all female species;* body size 5–7 cm; central Ariz. and southwestern N. Mex. southward into Mexico.

10b Enlarged preanal scales usually more than 3; tail light blue .... **Plateau Striped Whiptail,** *Cnemidophorus velox* **Springer**

Figure 360

Figure 360   Distribution of *C. velox*

Six or seven light stripes on dark brown or black; tail light blue, *all female species;* body size 6–8 cm. Distribution (Fig. 360).

11a (6) Circumorbital scales extend far forward, vertebral light stripe is wavy .... **New Mexico Whiptail,** *Cnemidophorus neomexicanus* **Lowe & Zweifel**

Seven light stripes on dark brown, diffuse spots on sides, tail greenish blue toward tip, *all female species;* body size 6–8 cm; central to southwestern N. Mex.

11b Circumorbital scales do not extend far forward (third supraocular in contact with frontal and frontoparietal scales); vertebral stripe, if present, not wavy **12**

12a Seven or eight longitudinal light stripes, spots between stripes confined to lateral dark fields **13**

12b Usually only six longitudinal light stripes, spots between central as well as lateral stripes **14**

13a Broad vertebral light stripe (sometimes divided into 2 stripes posteriorly) .... **Texas Spotted Whiptail,** *Cnemidophorus gularis* **Baird & Girard**

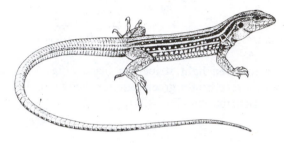

Figure 361

Figure 361   *Cnemidophorus gularis*

Seven or eight light stripes on dark brown with light spots on sides (Fig. 361); body size 6–9 cm; southern Oklahoma southward to Mexico.

13b Narrow vertebral light stripe .... **Laredo Striped Whiptail,** *Cnemidophorus laredoensis* **McKinney** *et al*

Color like *gularis, all female species;* body size 5–9 cm; restricted localities in southern Tex.

14a Two enlarged preanal scales, light spots less numerous rarely touching longitudinal stripes .... Gila Spotted Whiptail, *Cnemidophorus flagellicaudus* Lowe & Wright

Light lines and spots on dark brown, young without spots, *all female species;* body size 6–8.5 cm; northwestern Ariz. to southwestern N. Mex.

14b Three enlarged preanal scales, light spots more numerous often touching longitudinal stripes 15

15a Ratio of interparietal scale width/ length × 100 is greater than 65; young without light spots in dark fields between stripes .... Sonoran Spotted Whiptail, *Cnemidophorus sonorae* Lowe & Wright

Light stripes and spots on dark brown, young not spotted, tail reddish-tan; *all female species;* body size 6–8.5 cm; southwestern N. Mex. and southeastern Ariz. south into Mexico.

15b Ratio of interparietal scale width/ length × 100 is less than 65; young with light spots between lines .... Chihuahuan Spotted Whiptail, *Cnemidophorus exsanguis* Lowe

Light stripes and spots on dark brown, young spotted (not true of hatchlings), tail grayish-olive, *all female species;* body size 7–10 cm. Distribution (Fig. 362).

Figure 362

Figure 362   Distribution of *C. exsanguis*

SNAKES
Squamata, Serpentes

Of all the reptiles, snakes perhaps have held the greatest fascination and intrigue if not fear and misunderstanding by man. This is likely due to the fact that several groups of snakes are venomous which tend to give all species a bad name. Actually a majority of snakes are harmless and most, even the venomous species, are beneficial in consuming rodent pests.

Snakes have elongate bodies without functional limbs although a few primitive forms such as boas have vestigial pelvic girdles and a vestigial spur of a hind limb. The vertebrae are numerous and there is no sternum. The mandibles are connected by an elastic ligament which permits the jaw to expand and allows snakes to swallow large prey whole. There is a tendency for the internal organs to be elongated and the bilateral symmetry disrupted either by displacement or loss. For example the left lung in all snakes is either absent or rudimentary. Snakes have no moveable eyelid but the eye is covered by a transparent epidermal scale (brille) which is shed with the skin at periodic intervals. There is no external ear opening nor functional auditory apparatus. The tongue is forked and is used in conjunction with Jacobsen's Organ in olfaction and taste.

Snakes occur in a wide variety of terrestrial, aquatic, and arboreal habitats. Being poikilothermic they are more numerous in southern climates. Both oviparous and viviparous species exist. Snakes may eat insects, rodents, lizards, birds, fish, and a variety of other foods. Due to their secretive habits they are rarely encountered in large numbers except at "dens" where individuals of some species may form aggregations during winter months.

There are something over 2000 species in 13 families world-wide of which 119 species occur in the United States. Typical scale characteristics of snakes are given in Fig. 363.

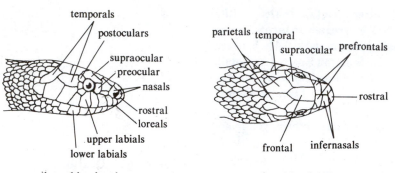

lateral head scales

dorsal head scales

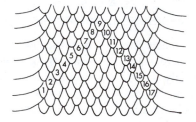

method of counting dorsal scales

Figure 363   Scales of snake and method of counting dorsal scale rows

KEY TO THE U.S. FAMILIES OF SNAKES

1a Belly scales transversely elongated and extending across the width of the venter (Fig. 364a) **2**

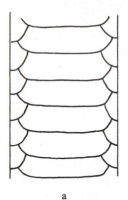

a   b

Figure 364   Ventral scales in typical snakes (a) and Leptotyphlops (b)

1b Belly scales not transversely elongated (Fig. 364b) ....
.... **(p. 168) SLENDER BLIND SNAKE FAMILY, Leptotyphlopidae**

2a A sensory pit between the eye and nostril on each side of the head (Fig. 365) ....
...... **(p. 171) VIPER FAMILY, Viperidae**

sensory pit

Figure 365   Sensory pit of vipers

2b No such sensory pit **3**

3a A pair of permanently erect fangs on the anterior part of the upper jaw...... **(p. 170) CORAL SNAKE FAMILY, Micruridae`**

3b No such permanently erect fangs........... **4**

4a Two or more pairs of chin shields (elongate, enlarged scales between right and left lower labials, Fig. 366a) ....
.... **(p. 179) COMMON SNAKE FAMILY, Colubridae**

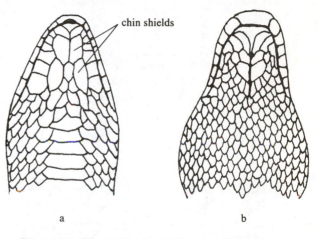

chin shields

a   b

Figure 366   Chin shields of colubrids (a) and boids (b)

4b No paired, enlarged chin shields (Fig. 366b)
....
.... **(p. 169) BOA and PYTHON FAMILY, Boidae**

SLENDER BLIND SNAKE FAMILY
Leptotyphlopidae

The slender blind snakes (also known as thread snakes) are small, wormlike in appearance and size. They are fossorial in habits and are rarely seen above ground except under flat rocks or occasionally moving between rocks in late evening. Blind snakes are unique in having belly scales like those of the dorsum rather than transverse scutes as in other snakes. Other characteristics include remnants of the pelvic girdle bones, reduced eyes which are covered by ordinary scales (no brille), teeth on the lower jaw only, and a short tail with a sharp point. Blind snakes are oviparous and feed largely on termites and ant larvae and eggs. Approximately 40 species in the single genus *Leptotyphlops* occur from the southwestern U.S. to South America and in Africa and southwestern Asia.

1a Supraoculars present (Fig. 367a) ....
.... Texas Blind Snake,
***Leptotyphlops dulcis* (Baird & Girard)**

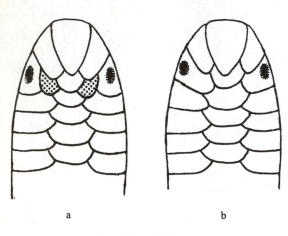

a   b

Figure 367

Figure 367   Supraoculars of *Leptotyphlops*

Uniform light brown to pinkish; total length 13–27 cm. Distribution (Fig. 368).

Figure 368

Figure 368   Distribution of *L. dulcis*

1b Supraoculars absent (Fig. 367b) ....
.... Western Blind Snake,
***Leptotyphlops humilis* (Baird & Girard)**

Uniform light brown to pinkish but usually with a distinctly lighter venter; total length 18–32 cm. Distribution (Fig. 369).

Figure 369

Figure 369   Distribution of *L. humilis*

BOA AND PYTHON FAMILY
Boidae

The Boidae include terrestrial and aboreal species which are heavy bodied with numerous, smooth, glossy scales. They kill their prey by constriction. The family is considered primitive and there is a vestigial pelvic girdle and rudimentary hind limbs which show externally as a spur on each side of the body near the vent. They are more common in tropical regions, but two species are native to the U.S.

1a Enlarged plates on top of head, 3 plates between eyes (central one largest) (Fig. 370a) .... Rubber Boa, *Charina bottae* **(Blainville)**

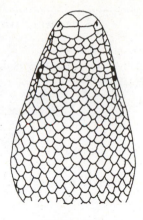

Figure 370b

Figure 370b *Lichanura trivirgata*

Color brown above with yellow venter; body length 35–75 cm. Distribution (Fig. 371).

Figure 370a

Figure 370a *Charina bottae*

Figure 371

Figure 371 Distribution of *C. bottae*

1b Top of head covered with small scales, several between eyes of equal size (Fig. 370b) .... Rosy Boa, *Lichanura trivirgata* **Cope**

Color gray, tan or rose with 3 broad longitudinal stripes (Fig. 372) or irregular brown patches; body length 60–100 cm; southern Calif. and southwestern Ariz., south into Sonora and Baja California, Mexico.

Figure 372

Figure 372   *Lichanura trivirgata*

CORAL SNAKE FAMILY
Micruridae

These snakes are relatively small and seldom bite but they have highly dangerous venom. There are two immovable hollow fangs in the front of the mouth. They are brightly colored with red, yellow and black bands and can be distinguished from their non-venomous mimics by the fact that the red and yellow bands are in contact. This does not hold for coral snakes world wide. The old adage referring to the bands, "red and yellow kill a fellow but red and black, venom is lack (or friend of Jack)" is useful only within the United States.

1a Red ring on neck...
.... Arizona Coral Snake,
Micruroides euryxanthus **(Kennicott)**

Color of red, yellow (sometimes whitish), and black rings, with black snout (red in kingsnake mimics); body length 35–53 cm; central to southeastern Ariz. and extreme southwestern N. Mex. and southward into Mexico.

1b Black ring on neck ....
.... Eastern Coral Snake,
Micrurus fulvius **(Linnaeus)**

Figure 373

Figure 373   *Micrurus fulvius*

Color of red, yellow and black rings, with black snout (Fig. 373); body length 50–100 cm. Distribution (Fig. 374).

Figure 374

Figure 374   Distribution of *M. fulvius*

VIPER FAMILY
Viperidae

This family comprises all the venomous snakes in the U.S. except coral snakes and includes copperheads, cottonmouths, and rattlesnakes. All new world representatives of this family have a facial pit (see below and Fig. 365) and are called Pit Vipers. Most old world vipers lack the facial pit. These species have an elaborate venom delivery mechanism consisting of a large hollow fang located anteriorly on each side of the upper jaw. The fangs are movable (erectile) and fold backward when the mouth is closed. These snakes are heavy bodied with rough scales and a broad triangular head. There is a temperature sensitive loreal pit organ between the eye and nostril on each side of the head. Rattlesnakes are so named because of the rattles at the end of the tail which produce a buzzing sound when the tail is vibrated. The rattles are a series of loosely connected horny segments. A new segment is produced each time the skin is shed (more frequently than once a year). Venom of pit vipers is generally hematoxic, whereas coral snake venom is neurotoxic.

1a Rattle at end of tail 2

1b No rattle at end of tail 3

2a Scales on top of head small (Fig. 375a)
.... 5

2b Scales on top of head consisting of enlarged plates (Fig. 375b) 4

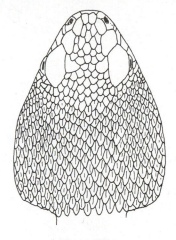

Figure 375a

Figure 375a   *Crotalus*

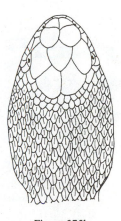

Figure 375b

Figure 375b   *Sistrurus*

3a A loreal scale present (Fig. 376a), dorsal scale rows fewer than 24 (usually 23) .... Copperhead, *Agkistrodon contortrix* (Linnaeus)

Color golden brown to reddish brown with darker cross bands (Fig. 377); body length 60–90 cm. Distribution (Fig. 378).

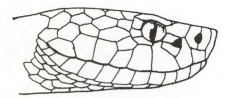

a

b

Figure 376

Figure 376   Loreal present (a) and absent (b)

Figure 377

Figure 377   *Agkistrodon contortrix*

Figure 378

Figure 378   Distribution of *A. contortrix*

3b Loreal scale absent (Fig. 376b), dorsal scales rows usually 25 ...... Cottonmouth, *Agkistrodon piscivorus* (Lacépède)

Olive, dark brown or black with dark crossbands sometimes obscured in darker individuals; body length 70–120 cm; southeastern U.S. from southern Va. to central Tex. including peninsular Fla. and extending up Mississippi Valley to central Mo.

4a Upper preocular not in contact with post-nasal (Fig. 379a) ...... Pigmy Rattlesnake, *Sistrurus miliarius* (Linnaeus)

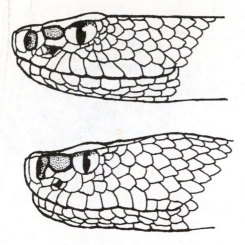

Figure 379

Figure 379   Postnasals of *Sistrurus*

Color gray to dark brown with dark dorsal blotches alternating with smaller lateral blotches; body length 35–55 cm; southeastern U.S. from southern N.C. to eastern Tex. and southern Mo. through peninsular Fla.

4b Upper preocular in contact with postnasal (Fig. 379b) .... Massasauga, *Sistrurus catenatus* (Rafinesque)

Color gray to brown with row of dark dorsal blotches alternating with double row of lateral blotches; body length 50–70 cm. Distribution (Fig. 380).

Figure 380

Figure 380   Distribution of *S. catenatus*

5a Supraoculars forming hornlike processes above eyes (Fig. 381) .... Sidewinder, *Crotalus cerastes* Hallowell

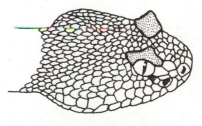

Figure 381

Figure 381   *Crotalus cerastes*

Dorsal color pale blending with prevailing soil color with various darker spots but without conspicuous patterns; body length 40–80 cm; southeastern Calif., southern Nev., and western Ariz.

5b No hornlike processes above eyes 6

6a Scales following dorsal contour of snout forming a prominent ridge (Fig. 382) .... Ridgenose Rattlesnake, *Crotalus willardi* Meek

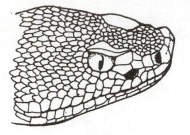

Figure 382

Figure 382 *Crotalus willardi*

Dorsum reddish brown to gray with short whitish crossbands edged with dark brown or black; body length 40–60 cm; rare; occurs in mountains of southeastern Ariz. and southwestern N. Mex. southward into Mexico.

6b Scales of snout not forming a ridge 7

7a Dorsal pattern of two parallel rows of small dark blotches (Fig. 383) .... Twin-spotted Rattlesnake, *Crotalus pricei* Van Denburgh

Figure 383

Figure 383 *Crotalus pricei*

Color grayish brown to gray with two rows of brown spots down its back; a small species, body length 20–50 cm; occurs in mountains of southeastern Ariz. southward into Mexico.

7b Dorsal pattern not as above 8

8a More than two internasals in contact with rostal (Fig. 384) .... Western Rattlesnake, *Crotalus viridis* (Rafinesque)

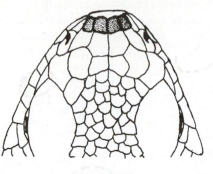

Figure 384

Figure 384   Four internasals contacting rostral

General color extremely variable from cream, yellow, gray, pink, brown or blackish with dark brown blotches on back; body length 40–150 cm, geographically variable. Distribution (Fig. 385).

Figure 385

Figure 385   Distribution of *C. viridis*

8b Only two internasals in contact with rostral 9

9a Dorsal pattern of widely spaced (about 8–9 scales between crossbands on midline) distinct dark crossbands (Fig. 386)................ .... Rock Rattlesnake, *Crotalus lepidus* (Kennicott)

Figure 386

Figure 386   *Crotalus lepidus*

Dorsal color gray often with bluish or greenish tint or tanish with widely and regularly spaced narrow black or dark brown crossbands faint anteriorly in eastern populations; body length 35–75 cm; distributed from southeastern Ariz. to southwestern Tex., southward into Mexico.

9b Dorsal pattern not as above 10

10a Dorsal pattern of many (often faint) cross-bands (about 2 scales between crossbands on midline) .... Tiger Rattlesnake, *Crotalus tigris* Kennicott

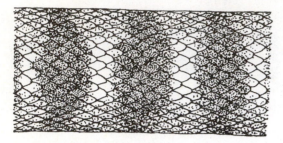

Figure 387

Figure 387 *Crotalus tigris*

Color gray, pink, or tan with numerous irregular crossbands (Fig. 387) of gray or brown with faint borders; body length 45–90 cm; southern Ariz., southward into Sonora, Mexico.

10b Dorsal pattern not as above 11

11a Dorsal pattern with salt and pepper speck-ling, prenasal separated from rostral by small scales, (Fig. 388) (southwestern sub-species) or supraoculars divided, pitted, or creased (Panamint subspecies) .... Speckled Rattlesnake, *Crotalus mitchelli* (Cope)

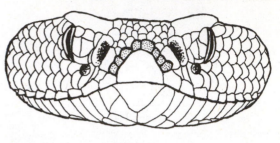

Figure 388

Figure 388 *C. mitchelli*

Dorsal color variable from cream, tan, yellow, pink, gray or brown harmonizing with substrate, pattern indistinct and variable but with salt-and-pepper speckling; body length 60–130 cm; south-ern Calif., southern Nev., and western Ariz.

11b Dorsal pattern not speckled, prenasal not separated by small scales 12

12a Tail with distinct black and white rings 13

12b Tail without rings or faintly ringed with light rings similar to ground color of body 15

13a Light tail rings distinctly broader than dark rings (Fig. 389) ...... Mohave Rattlesnake, *Crotalus scutulatus* **Kennicott**

Figure 389

Figure 389 *Crotalus scutulatus*

Color greenish gray to olive brown with distinct dark blotches, diamonds, or hexagons down middle of back, tail ringed black and white; body length 60–130 cm; from eastern Calif. and southern Nev. across southwestern Ariz. into Mexico and southwestern Tex.

13b Light and dark tail rings approximately equal in width 14

14a Color red or reddish brown, first pair of lower labials usually divided transversely (Fig. 390a) ... Red Diamond Rattlesnake, *Crotalus ruber* **Cope**

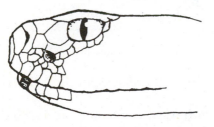

Figure 390a

Figure 390a *C. ruber*

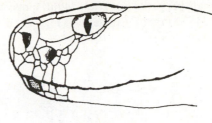

Figure 390b

Figure 390b *C. atrox*

Color reddish with dorsal diamond blotches less well defined than Western Diamondback vicar and without or with only faint pepper markings; body length 75–165 cm; Baja California entering U.S. only in extreme southwestern Calif.

14b Color gray or brown, first pair of lower labials not divided transversely (Fig. 390b) .... Western Diamondback Rattlesnake, *Crotalus atrox* **Baird and Girard**

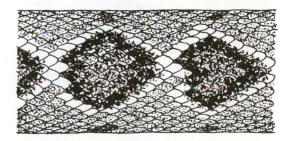

Figure 391

Figure 391 *C. atrox*

Dorsum gray to brown with brown diamonds (Fig. 391) or hexagonal blotches often made indefinite with small peppered dark and light spots; body length 75–225 cm. Distribution (Fig. 392).

Figure 392

Figure 392   Distribution of *C. atrox*

**15a Dorsal markings consisting of distinctly outlined diamond shapes (Fig. 393) ....
.... Eastern Diamondback Rattlesnake, *Crotalus adamanteus* Beauvois**

Figure 393

Figure 393   *Crotalus adamanteus*

Olive, brown or blackish with dark brown or black diamonds along back; body length 85–240 cm. Distribution (Fig. 394).

Figure 394

Figure 394   Distribution of *C. adamanteus*

15b Dorsal pattern of dark blotches or cross-bands; tail is black 16

16a Black of tail ending abruptly at body, white blotches in dark dorsal crossbands (Fig. 395) .... Blacktail Rattlesnake, *Crotalus molossus* Baird & Girard

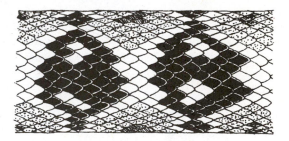
Figure 395

Figure 395   *Crotalus molossus*

Dorsum gray, olive or brown with dark brown or black crossbands containing light blotches; body length 70–130 cm; central Tex. to central and southwestern Ariz. southward into Mexico.

16b Black of tail usually grades into body coloration, distinct dark dorsal crossbands or blotches without light markings ....
.... Timber Rattlesnake,
Crotalus horridus Linnaeus

Color variable from almost totally black to blackish, brown, gray, tan, yellow with dark blotches or crossbands; body length 90–190 cm. Distribution (Fig. 396).

Figure 396

Figure 396   Distribution of *C. horridus*

COMMON SNAKE FAMILY
Colubridae

The Colubridae is the largest family of snakes comprising over 75% of all snakes in the world and almost 85% of the U.S. snakes. Recent students of snake systematics suggest that the family may be polyphyletic thus it eventually may be divided into several families. Colubrids are as diverse in their structure and habits as they are in their numbers. Their body form and size are extremely variable from long-thin to short-robust forms both large (over 7 feet in U.S. species of *Elaphe, Masticophis,* and *Pituophis*) and small (several inches in *Virginia*). U.S. forms are not venomous to man and most are beneficial but a few have evolved a rear-fanged venom apparatus used in capturing prey. Rear-fanged snakes, those with modified teeth in the rear of the mouth and venom glands include species of *Coniophanes, Trimorphodon,* and *Leptodeira. Hypsiglena* has venom but no tooth grooves or specialized fang apparatus whereas *Tantilla, Ficimia* and *Gyalopion* have grooved fangs but no venom.

Colubrid snakes may be fossorial, aquatic, terrestrial, or arboreal. Their locomotion is entirely "serpentine" aided by broad transverse ventral scales. There are no vestiges of limbs or girdle bones. Food habits are carnivorous but their diets are varied from worms and insects in smaller

forms to birds and mammals in larger species. Aquatic species consume fishes and amphibians. Colubrids occur in a variety of habitats and usually have smaller population densities than lizards although the habit in some species of aggregating for winter hibernation may result in concentrated populations (e.g., several thousand garter snakes in a winter hibernaculum).

Generally the colubrid snakes are very secretive. A common technique employed to hunt snakes involves driving along country roads at night just after dusk or after a summer rain. Many species are attracted to the warm pavement in the early evening.

Colubrids may be oviparous, ovoviviparous, or viviparous. Aquatic species tend to be viviparous. Clutch or litter size varies with body size but may average near 50 eggs in some species. Some species such as *Farancia* guard their eggs during the incubation period.

As with other reptiles the shape, size, number, and kinds of scales are commonly used for identification along with color pattern. Nomenclature of typical scale patterns is given in Fig. 363.

1a Dorsal scales are keeled (Fig. 397a) 2

Note: Some species have weakly keeled scales not strikingly evident in some specimens (e.g., *Elaphe*) and some are predominantly smooth with keels occasionally in males (e.g., *Phyllorhynchus*) or on tail *(Regina alleni)* or have lines on lateral scales which appear to be keels *(Seminatrix).*

1b Dorsal scales are smooth (Fig. 397b)... 48

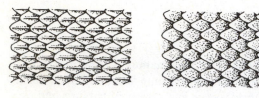

Figure 397

Figure 397   Dorsal scales keeled (a) or smooth (b)

2a Anal plate divided (Fig. 398a) 3

2b Anal plate not divided (Fig. 398b) 31

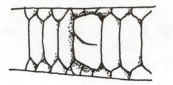

a

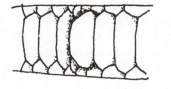

b

Figure 398

Figure 398   Anal plate divided (a) or not divided (b)

3a Rostral turned upward and keeled (Fig. 399) 4

Figure 399

Figure 399   Rostral of *Heterodon*

3b Rostral not turned upward 6

4a Underside of belly and tail is black with white or yellow patches .... Western Hognose Snake, *Heterodon nasicus* Baird and Girard

Figure 400

Figure 400 *Heterodon nasicus*

Dorsum light brown with rows of dark brown spots, large brown blotch on sides of neck (Fig. 400); body length 40–80 cm. As with other hognose snakes, exhibits the behavior of "playing dead" by rolling onto back when captured. Distribution (Fig. 401).

Figure 401

Figure 401 Distribution of *H. nasicus*

4b Underside of belly not black, may be white, light gray, or light green, and may have mottled appearance of gray on lighter yellow or pink 5

5a Underside of tail lighter than belly .... Eastern Hognose Snake, *Heterodon platyrhinos* Latreille

Dorsal color variable from total black but usually with large brown spots and light bands on brown, olive, or reddish background; body length 50–90 cm. Distribution (Fig. 402).

Figure 402

Figure 402 Distribution of *H. platyrhinos*

5b Underside of tail similar to body .... Southern Hognose Snake, *Heterodon simus* (Linnaeus)

Dorsal color similar to western hognose except spots tend to be larger and darker; body length 35–55 cm. Distribution (Fig. 403).

Figure 403

Figure 403 Distribution of *H. simus*

6a No loreal scale, preocular in contact with eye (Fig. 404a) 7

6b Loreal scale present (Fig. 404b) 8

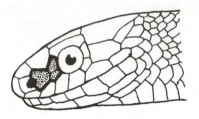

a

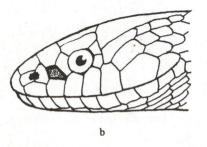

b

Figure 404

Figure 404   Loreal scale absent (a) or present (b)

7a Dorsal scales in 17 rows, venter pale yellow, pink or light brown .... Brown Snake, *Storeria dekayi* (Holbrook)

Dorsum shades of brown (reddish, yellowish to grayish) with double row of black spots down back, four subspecies in eastern U.S. with various color variations including connecting lines between spots and dark ring on neck; body length 20–45 cm. Distribution (Fig. 405).

Figure 405

Figure 405   Distribution of *Storeria dekayi*

7b Dorsal scales in 15 rows, venter red or red-orange .... Redbelly Snake, *Storeria occipitomaculata* (Storer)

Figure 406

Figure 406   *Storeria occipitomaculata*

Color brown to gray with 4 narrow dark stripes and/or a light middorsal stripe, head darker with 3 light spots on neck (Fig. 406), belly (normally red) may vary from yellow to occasional blue-black variants; body length 20–40 cm. Distribution (Fig. 407).

Figure 407

Figure 407   Distribution of *S. occipitomaculata*

8a 1 internasal (Fig. 408a) **9**

8b 2 internasals (Fig. 408b)....................... **12**

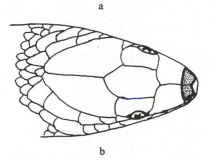

a

b

Figure 408

Figure 408   One (a) or two (b) internasals

9a 5–6 upper labials (Fig. 409a), 15–17 dorsal scale rows **10**

9b 7–8 upper labials (Fig. 409b), 19–21 dorsal scale rows **11**

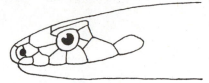

a

b

Figure 409

Figure 409   Upper labials six (a) or seven (b)

10a 5 upper labials, dorsals strongly keeled .... **Rough Earth Snake, *Virginia striatula* (Linnaeus)**

Plain brown (grayish to reddish) dorsum; venter yellow to pink; body length 18–30 cm. Distribution (Fig. 410).

Figure 410

Figure 410   Distribution of *Virginia striatula*

10b 6 upper labials, dorsals weakly keeled or smooth .... **Smooth Earth Snake,** *Virginia valeriae* **Baird & Girard**

Color uniform gray to reddish brown with white belly, eastern subspecies may have tiny black spots on dorsum; body length 18–30 cm. Distribution (Fig. 411).

Figure 411

Figure 411   Distribution of *V. valeriae*

11a No preocular, dorsal scales predominantly smooth, keeled only in supra-anal region **101**

11b 1 or 2 preoculars, dorsal scales predominantly smooth, keeled only in anal region and on top of tail .... **Striped Crayfish Snake,** *Regina alleni* **(Garman)**

Figure 412

Figure 412   *Regina alleni*

Broad dark brown stripe down back with broad yellow stripe on sides (Fig. 412), venter yellowish with or without dark midventral markings; body length 30–60 cm; found in peninsular Fla. to southern Ga.

12a No preoculars (loreal in contact with eye) **13**

12b 1 or 2 preoculars **14**

13a 5–6 upper labials .... *Virginia* **(see couplet number 10)**

13b 7 upper labials **101**

14a 17 scale rows **15**

14b More than 17 scale rows **16**

15a 7 upper labials, 7 or 8 lower labials .... **Rough Green Snake,** *Opheodrys aestivus* **(Linnaeus)**

Uniform light green dorsum with white, yellow or greenish venter; slender body 55 to 115 cm in length. Distribution (Fig. 413).

Figure 413

Figure 413   Distribution of *Opheodrys aestivus*

15b 9 upper labials, 10 or 11 lower labials .... **Speckled Racer,** *Drymobius margaritiferus* **(Schlegel)**

Dorsum bluish-black with yellow spot in center of each dorsal scale, belly plain white or yellow; body length 75–125 cm; a Middle American species entering U.S. only in Rio Grande valley of extreme southern Tex.

16a 3 postoculars, or if 2 postoculars then 19–23 scale rows **17**

16b 2 postoculars, 25 or more scale rows (weakly keeled) **27**

17a 19 scale rows **18**

17b More than 19 scale rows **21**

18a Dorsum of large dark spots over reddish brown, venter red bordered by rows of black spots .... **Kirtland's Snake,** *Clonophis kirtlandi* **(Kennicott)**

Dorsum reddish brown with 2 rows of dark spots on each side of body; body length 35–55 cm. Distribution (Fig. 414).

Figure 414

Figure 414   Distribution of *Clonophis kirtlandi*

18b Dorsum generally uniform dark with lateral light stripe, venter not red **19**

19a Venter uniform light or with single stripe down center .... **Graham's Crayfish Snake,** *Regina grahami* **Baird & Girard**

Dark brown dorsum with broad yellow to cream colored stripe on sides bordered ventrally with dark stripe; body length 45–105 cm. Distribution (Fig. 415).

Figure 415

Figure 415   Distribution of *Regina grahami*

19b Venter with more than single dark stripe **20**

20a Venter with 2 dark stripes down center and a dark stripe on outer edge of ventral scales (Fig. 416a) .... Queen Snake, *Regina septemvittata* (Say)

20b Venter with various dark markings sometimes resembling stripes but not as above (Fig. 416b)....................................... .... Glossy Crayfish Snake, *Regina rigida* (Say)

Figure 416a

Figure 416a *Regina septemvittata*

Figure 416b

Figure 416b *Regina rigida*

Dorsum like *R. grahami* except lateral stripe narrower and venter with 2 dark stripes; body length 35–85 cm. Distribution (Fig. 417).

Shiny dorsum of plain brown or with faint dark stripes, very narrow light stripe on side; body length 35–75 cm. Distribution (Fig. 418).

Figure 418

Figure 418   Distribution of *R. rigida*

Figure 417

Figure 417   Distribution of *R. septemvittata*

21a 1 or 2 suboculars (Fig. 419a), upper labial scales not reaching eye.. Green Water Snake, *Nerodia cyclopion* (Dumeril, Bibron, & Dumeril)

a

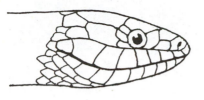

b

Figure 419

Figure 419   Suboculars present (a) or absent (b)

Coloration greenish or brownish with dark irregular markings and/or light spots, no distinct pattern; venter plain white or cream in Fla. subspecies or grayish with yellow spots in western subspecies; body length 75–180 cm. Distribution (Fig. 420).

Figure 420

Figure 420   Distribution of *N. cyclopion*

21b No suboculars (upper labials contact eye, Fig. 419b) 22

22a Dorsal pattern of 4 rows of dark spots, belly with distinct row of small dark spots on each side .... Harter's Water Snake, *Nerodia harteri* (Trapido)

Color reddish brown with 4 rows of darker spots, belly pink or orange with dark spots; body length 50–80 cm; two subspecies with restricted distributions in central Tex., one in Brazos River system and other in Concho-Colorado River system.

22b Color pattern not as above 23

23a 27 or more dorsal scale rows, 11–13 lower labials 24

23b 21–25 (rarely 27) dorsal scale rows, usually 10 lower labials 25

24a Dorsal pattern of square dark blotches on lighter brown, middorsal blotches separate from lateral blotches (Fig. 421)................. .... Brown Water Snake, *Nerodia taxispilota* (Holbrook)

Figure 421

Figure 421   *Nerodia taxispilota*

Belly yellow and brown with spots and half moons of black; body length 75–165 cm; eastern Coastal Plain from Va. to Ala. and throughout Fla.

24b Dorsal pattern with lighter middorsal and lateral blotches (often diamond or hexagonal) on darker background which forms a chainlike pattern on dorsum .... Diamondback Water Snake, *Nerodia rhombifera* **(Hallowell)**

Belly predominantly yellow with small dark spots concentrated near sides; body length 75–150 cm. Distribution (Fig. 422).

Figure 422

Figure 422   Distribution of *N. rhombifera*

25a Dark crossbands on neck and anterior (approximate 1/4) of body and alternating dark dorsal and lateral blotches on rest of body .... Northern Water Snake, *Nerodia sipedon* **(Linnaeus)**

Brown or reddish brown with large dark blotches forming crossbands on anterior of body, venter variable from yellow to brown with brown or reddish crescents or half moons; body length 55–130 cm. Distribution (Fig. 423).

Figure 423

Figure 423   Distribution of *N. sipedon*

25b Pattern not as above 26

26a Venter plain (yellowish or red) or with bases of belly scales slightly dark on sides (Fig. 424a), dorsal pattern uniform reddish brown or greenish gray or with distinct middorsal row of large blotches alternating with smaller lateral blotches .... Plainbelly Water Snake, *Nerodia erythrogaster* (Forster)

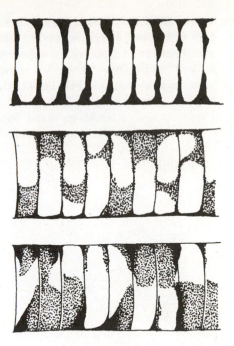

Figure 424b

Figure 424b   Venter variation in *N. fasciata*

Body length 76–140 cm. Distribution (Fig. 425).

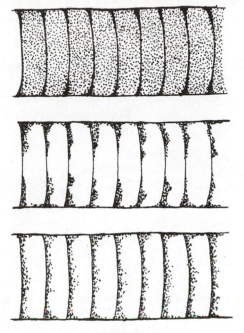

Figure 424a

Figure 424a   Venter variation in *N. erythrogaster*

Figure 425

Figure 425   Distribution of *N. erythrogaster*

26b Venter not plain*, with darker squares or dark with row(s) of light spots (Fig. 424b), dorsal pattern variable with crossbands or distinct light stripes (*Note: Red phase of Mangrove subspecies in Fla., belly tends to be plain as is dorsum) .... Southern Water Snake, *Nerodia fasciata* (Linnaeus)

Body length 60–150 cm. Distribution (Fig. 426).

Figure 426

Figure 426   Distribution of *N. fasciata*

27a (16) Suboculars present .... Trans-Pecos Rat Snake, *Elaphe subocularis* (Brown)

Figure 427

Figure 427   *Elaphe subocularis*

Dorsum tan to olive-yellow with dark blotches forming a double dark line on anterior of body (Fig. 427); body length 90–160 cm; distributed from southern N. Mex. and western Tex. southward into Mexico.

27b Suboculars absent.................................. 28

28a Dorsum uniform olive to pale green .... Green Rat Snake, *Elaphe triaspis* Cope

Color olive or pale green with white, cream or yellowish belly; body length 60–125 cm; found in western Mexico and enters U.S. only in southeastern Ariz.

28b Dorsum black or with blotches or stripes 29

29a Neck bands crossing parietal uniting on frontal to form "V" on top of head (Fig. 428)... Corn Snake *Elaphe guttata* (Linnaeus)

Figure 428

Figure 428   *Elaphe guttata*

Dorsum grayish to yellowish brown with brown or reddish blotches, belly checkered with black on white; body length 75–175 cm. Distribution (Fig. 429).

Figure 429

Figure 429   Distribution of *E. guttata*

29b No neck bands forming "V" as above **30**

30a More than 220 ventrals .... **Rat Snake,** *Elaphe obsoleta* **(Say)**

Dorsal pattern extremely variable in the 6 U.S. subspecies from plain black to dark stripes on yellowish, greenish or reddish or dark blotches on yellowish or grayish background; body length 80–215 cm. Distribution (Fig. 430).

Figure 430

Figure 430   Distribution of *E. obsoleta*

30b Less than 220 ventrals .... **Fox Snake,** *Elaphe vulpina* **(Baird & Girard)**

Dorsum yellowish brown or tan with middorsal and lateral rows of dark brown spots and blotches; body length 90–175 cm. Distribution (Fig. 431).

Figure 431

Figure 431   Distribution of *E. vulpina*

31a (2) 27 or more scale rows, usually 4 prefrontals **32**

31b Less than 27 scale rows, 2 prefrontals **33**

32a Dark line from eye to angle of jaw (Fig. 432) .... **Gopher Snake,** *Pituophis catenifer* **(Blainville)**

Figure 432

Figure 432   *Pituophis catenifer*

Color tan or yellowish with dark or reddish brown to black blotches, belly yellow with black spots; body length 125–185 cm. Some authorities believe that *P. catenifer* and *P. melanoleucas* should be combined based on tenuous evidence from eastern Tex. Distribution (Fig. 433).

Figure 433

Figure 433   Distribution of *P. catenifer*

32b Head black or, if light, no dark line from eye to angle of jaw .... Pine Snake, *Pituophis melanoleucas* (Daudin)

Dorsal color variable white, yellowish or tan with diffuse brown or distinct black blotches or dorsum uniform black, venter white with black blotches or black with or without white blotches; body length 120–170 cm. Distribution (Fig. 434). Note: Some authors recognize only *P. melanoleucas* (combining *P. catenifer* here).

Figure 434

Figure 434   Distribution of *P. melanoleucas*

33a Rostral very large (Fig. 435), scales weakly keeled and only in males 52

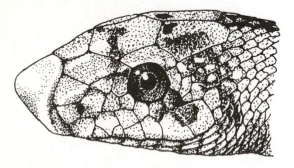

Figure 435

Figure 435   *Phyllorhynchus* showing large rostral

33b Rostral normal 34

34a Less than 8 lower labials 35

34b 8 or more lower labials 36

35a Two rows of black belly spots (often resembling half moons), dorsal pattern of lines .... Lined Snake, *Tropidoclonion lineatum* (Hallowell)

Dorsum brown with thin middorsal stripe of cream, tan, or orange and broad lateral light stripes on either side on a thin dark stripe; body length 20–55 cm. Distribution (Fig. 436).

Figure 436

Figure 436   Distribution of *T. lineatum*

35b Dorsum and venter plain .... (see 10a) *Virginia striatula* (Linnaeus)

36a No lateral light stripe 37

36b Lateral light stripe present 38

37a Middorsal stripe present (Fig. 437) ....
.... Northwestern Garter Snake,
Thamnophis ordinoides
(Baird & Girard)

Figure 437

Figure 437   *Thamnophis ordinoides*

Color brown with no or occasionally faint lateral light stripe and middorsal stripe of yellow, orange, or red occasionally faint or absent, venter yellow, olive, or gray with red or black blotches; body length 35–65 cm; distributed from extreme northwestern Calif. north along Pacific coast to southwestern British Columbia.

37b No middorsal stripe, dorsal pattern of
spots ....
.... Narrowhead Garter Snake,
Thamnophis rufipunctata **(Cope)**

Dorsum olive or brown with dark brown spots, no stripes; body length 50–80 cm. Distribution (Fig. 438).

Figure 438

Figure 438   Distribution of *T. rufipunctata*

38a Lateral stripe restricted to 3rd scale row
on neck, on rows 2 and 3 farther back ....
.... Checkered Garter Snake,
Thamnophis marcianus
(Baird & Girard)

Figure 439

Figure 439   *Thamnophis marcianus*

Dorsal color of black checkerboard pattern on olive tending to obscure middorsal and lateral stripes (Fig. 439); body length 45–80 cm. Distribution (Fig. 440).

Figure 440

Figure 440   Distribution of *T. marcianus*

38b Lateral stripe involving at least two scale
rows on anterior part of body 39

39a Lateral stripe involving 4th scale row, usu-
ally rows 3–4 (occasionally 2–4) 40

39b Lateral stripe not involving 4th scale row,
usually rows 2–3 (occasionally 1–3) (*foot-
note) 44

*Footnote: *T. brachystoma* occasionally with lateral stripe on lower part of 4th otherwise only on row 2 and 3.

40a Lateral stripe involving second scale row .... **Butler's Garter Snake,**
Thamnophis butleri (Cope)

Color olive-brown to black, stripes yellow to cream or lateral stripes may tend to orange; body length 35–55 cm; distributed from eastern Ind. and western Ohio northward through eastern Mich. and tip of southern Ontario, a disjunct population in southeastern Wisc.

40b Lateral stripe not involving second scale row **41**

41a Paired black blotches behind head .... **Mexican Garter Snake,**
Thamnophis eques (Reuss)

Dorsum brown to olive with yellowish middorsal and lateral stripe, sides checkered with dark spots; body length 45–100 cm; distributed from central Ariz. southeastward into Mexico and up Gila River into N. Mex.

41b No paired black blotches immediately behind head **42**

42a Upper labials boldly marked with dark bars on edges .... **Plains Garter Snake,**
Thamnophis radix (Baird & Girard)

Dorsum brown to reddish with black spots between stripes, middorsal stripe yellow or orange, lateral stripe yellow to green or bluish, belly light with spots on sides; body length 50–80 cm. Distribution (Fig. 441).

Figure 441

Figure 441   Distribution of *T. radix*

42b Upper labials immaculate without dark markings **43**

43a Dark ventrolateral stripe involving rows 1 and 2 and outer edge of belly, light spots on parietal of head faint or lacking .... Eastern Ribbon Snake, *Thamnophis sauritus* (Linnaeus)

Brown to black with cream or yellowish stripes, one subspecies of northwestern coast of peninsular Fla. with blue stripes; body length 45–85 cm and very slender. Distribution (Fig. 442).

Figure 442

Figure 442   Distribution of *T. sauritus*

43b No dark ventrolateral stripe, light spot on parietals of head prominent and touch each other .... Western Ribbon Snake, *Thamnophis proximus* (Say)

Color similar to Eastern Ribbon Snake except tends to have light flecks in dark fields between lines and middorsal stripe tends to be orange or red; body length 50–90 cm. Distribution (Fig. 443).

Figure 443

Figure 443   Distribution of *T. proximus*

44a (39) 6 upper labials, 17 dorsal scale rows .... Shorthead Garter Snake, *Thamnophis brachystoma* (Cope)

Color brown with cream to tan stripes, very small head no wider than neck and shorter than other garter snakes; body length 35–50 cm; found in Allegheny Plateau of northwestern Pa. and extreme southwestern N.Y.

44b 7–8 upper labials, more than 17 dorsal scale rows 45

45a 7 upper labials ... Common Garter Snake,
Thamnophis sirtalis **(Linnaeus)**

Most widespread and variable garter snake, brown, olive, or blackish with stripes or dark spots predominating, stripes may be yellow, tan, greenish or bluish and usually with double row of alternating black spots between stripes; body length 45–120 cm. Distribution (Fig. 444).

Figure 444

Figure 444   Distribution of *T. sirtalis*

45b 8 upper labials 46

46a Paired black blotches behind head ....
.... Blackneck Garter Snake,
Thamnophis cyrtopsis **(Kennicott)**

Dorsum with checkerboard pattern of black squares on olive background with orange mid-dorsal stripe and yellow lateral stripe (western subspecies) or tending to have checkered pattern of enlarged black or brown blotches on olive to tan with orange middorsal stripe and lateral stripe obscured or absent (eastern subspecies); body length 40–90 cm. Distribution (Fig. 445).

Figure 445

Figure 445   Distribution of *T. cyrtopsis*

46b No paired black blotches behind head
.. 47

47a Internasals wider than long and blunt anteriorly (Fig. 446a) ....
.... Western Terrestrial Garter Snake, *Thamnophis elegans* (Baird & Girard)

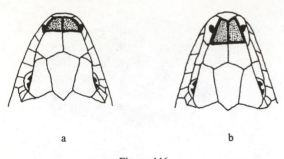

Figure 446

Figure 446 Internasals wide (a) or long (b)

Considerable variation in color pattern, brown, olive to gray dorsum with middorsal stripe white to orange and alternating double row of dark spots between lines tending to be rounded; body length 45–90 cm. Distribution (Fig. 447).

Figure 447

Figure 447 Distribution of *T. elegans*

47b Internasals longer than wide and pointed anteriorly (Fig. 446b).....................
.... Western Aquatic Garter Snake, *Thamnophis couchi* (Kennicott)

Color pattern highly variable, dark brown or olive with stripes to reddish brown with checkerboard pattern of dark blotches between stripes; body length 45–130 cm; from southwestern Ore. through most of Calif. except southeast, south to northern Baja California.

48a (1) Anal plate divided 62

48b Anal plate not divided.......................... 49

49a All or most subcaudals divided (Fig. 448a)
.... 50

49b Most subcaudals not divided (Fig. 448b)
.... Longnose Snake,
Rhinocheilus lecontei (Baird & Girard)

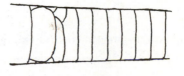

a

b

Figure 448

Figure 448 Subcaudals divided (a) or not divided (b)

Figure 449

Figure 449   *Rhinocheilus lecontei*

Color pattern of alternating red and black cross-bands, red bands speckled with black and black bands speckled with yellow and tending to be bordered with yellowish lines (Fig. 449); body length 55–100 cm. Distribution (Fig. 450).

Figure 450

Figure 450   Distribution of *R. lecontei*

50a Pupils vertical **51**

50b Pupils round **53**

51a Suboculars present, rostral enlarged, leaf shaped (Fig. 435) **52**

51b Suboculars absent, rostral not enlarged .... **Lyre Snake,** *Trimorphodon biscutatus* **(Dumeril, Bibron & Dumeril)**

Figure 451

Figure 451   *Trimorphodon biscutatus*

Body tan, yellowish or gray with dark brown saddles distributed along length, a lyre-shaped mark on head (Fig. 451), which may be faint or absent in Tex. subspecies; body length 45–100 cm. Distribution (Fig. 452).

Figure 452

Figure 452   Distribution of *T. biscutatus*

52a (33) Dorsal color pattern with small dark spots or blotches (Fig. 453) .... **Spotted Leafnose Snake,** *Phyllorhynchus decurtatus* **(Cope)**

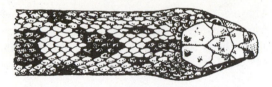

Figure 453

Figure 453   *Phyllorhynchus decurtatus*

Dorsum tan to light gray or tinged with pink or yellow with brown blotches or spots; body length 30–50 cm; found in southwestern Ariz., southern Nev. and southeastern Calif. southward in Mexico.

52b Dorsal color pattern of large dark saddles giving banded appearance (Fig. 454) .... **Saddled Leafnose Snake,** *Phyllorhynchus browni* **Stejneger**

Figure 454

Figure 454   *Phyllorhynchus browni*

Dorsal pattern of dark brown bands alternating pink or cream thinner bands; body length 30–50 cm; found in south central Ariz. barely extending into Mexico near Organ Pipe Cactus National Monument.

53a Loreal scale absent... Short-tailed Snake, *Stilosoma extenuatum* **Brown**

Color pattern of central row of small dark brown to black blotches and smaller blotches on sides separated with areas of red, orange, or yellow, venter with brown or black blotches; body length 35–65 and slender; restricted to west-central Fla.

53b Loreal scale present **54**

54a Venter light without dark markings **55**

54b Venter with at least some dark markings **56**

55a Upper labials 6–7, lower labials 8 .... **Scarlet Snake,** *Cemophora coccinea* **(Blumenbach)**

Dorsal pattern of red saddles bordered by black separated by yellowish area; body length 35–75 cm. Distribution (Fig. 455).

Figure 455

Figure 455   Distribution of *C. coccinea*

55b Upper labials 8, lower labials 12–15
.... **Glossy Snake,**
Arizona elegans **Kennicott**

Shiny snake with light brown blotches on cream or buff background; body length 65–135 cm. Distribution (Fig. 456).

Figure 456

Figure 456   Distribution of *Arizona elegans*

56a Dorsal scale rows 17 .... Indigo Snake,
Drymarchon corais **(Daudin)**

Color pattern shiny bluish black including venter, eastern subspecies tending to have reddish brown skin and Tex. subspecies tending to have anterior body brownish with hint of dark blotched pattern and dark stripe extending down from eye; body length 150–250 cm; two subspecies one in south Tex. extending southward into Middle America and second one in southern Ga., Fla. with isolated population in southern Ala.

56b Dorsal scale rows more than 17........... 57

57a Dorsal pattern of white bordered gray crossbands alternating with black bordered red-orange blotches on bands ....
.... **Gray-banded Kingsnake,**
Lampropeltis mexicana **(Garman)**

Above color pattern variable from light to dark gray and wide to narrow red-orange blotches or bands; body length 50–120 cm; found in southwestern Tex. southward into Mexico.

57b Dorsal pattern not as above 58

58a Dorsal pattern predominantly with small yellow spots forming "salt and pepper" appearance (Fig. 457a) or chain pattern (Fig. 457b) or predominantly of dark irregular rings alternating with white or cream colored rings (Fig. 458a) or striped (Fig. 458b) .... **Common Kingsnake,**
Lampropeltis getulus **(Linnaeus)**

a

b

Figure 457

Figure 457   *Lampropeltis getulus* pattern variation (east)

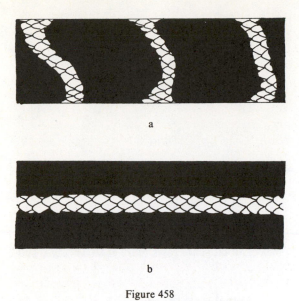

a

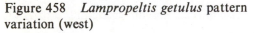

b

Figure 458

Figure 458   *Lampropeltis getulus* pattern variation (west)

Color pattern extremely variable; body length 90–200 cm. Seven subspecies often with distinct color pattern as in Fig. 457 (eastern part of range) and 458 (western part of range). Distribution (Fig. 459).

Figure 459

Figure 459   Distribution of *L. getulus*

58b Dorsal pattern not as above **59**

59a 25–27 rows of scales at midbody *and* fewer than 58 subcaudal scales ....
.... **Prairie Kingsnake,**
Lampropeltis calligaster **(Harlan)**

Dorsal pattern of brown or reddish blotches with black edges and double row of blotches on side (fuse into one blotch in some specimens), eastern subspecies tends to have only faint blotches and often appears uniform brown; body length 75–130 cm. Distribution (Fig. 460).

Figure 460

Figure 460   Distribution of *L. calligaster*

59b 19–23 scale rows at midbody, if 25 rows then 59–79 subcaudal scales **60**

60a Dorsal pattern in rings of red, black and white, end of snout uniform white; scale rows at midbody 23–25 ....
.... **Sonoran Mountain Kingsnake,**
Lampropeltis pyromelana **(Cope)**

Body length 45–105 cm; found in mountains in central southwestern Utah and central to southeastern Ariz. and southward into Mexico.

60b If dorsal pattern in rings end of snout not uniform white; scale rows at midbody 19–23 .... **61**

61a Pattern of red, black and white rings with black snout occasionally with red markings ..... California Mountain Kingsnake, *Lampropeltis zonata* (Lockington)

Body length 50–100 cm; found in mountains from northern Baja through coastal, central and northern Calif. and southwestern Oreg. and an isolated population in southern Wash.

61b Pattern of broad red and black bands, and white or yellow rings (Fig. 461a), or in eastern subspecies pattern of large reddish brown blotches with a Y or V shaped mark (Fig. 461b) in blotch behind head (resembles Prairie Kingsnake which doesn't have this mark and has smaller blotches) .... Milk Snake, *Lampropeltis triangulum* (Lacépède)

a

b

Figure 461

Figure 461   Pattern variations in *L. triangulum*

Body length 36–130 cm. Scale rows at midbody 19–23. Distribution (Fig. 462).

Figure 462

Figure 462   Distribution of *L. triangulum*

62a (48) Dorsal scale rows less than 19 ..... **63**

62b Dorsal scale rows more than 19......... **100**

63a Loreal scale present **64**

63b Loreal scale absent **82**

64a No preoculars **65**

64b Preoculars present.............................. **66**

65a Dorsal scale rows 13, 5 upper labials .... Worm Snake, *Carphophis amoenus* (Say)

Uniform brown or black back with pink belly; body length 19–35 cm. Distribution (Fig. 463).

Figure 463

Figure 463   Distribution of *Carphophis amoenus*

65b More than 13 dorsal scale rows, 6 upper labials .... (see couplet 10b)
Virginia valeriae Baird & Girard

66a A single preocular **67**

66b Preoculars 2 or 3................................. **73**

67a Dorsal scale rows 17 .... **68**

67b Dorsal scale rows less than 17 .... **69**

68a Dorsum black, venter red ....
.... **Black Swamp Snake,**
Seminatrix pygaea (Cope)

Shiny black dorsum with red venter; body length 25–45 cm; distributed through Fla. to southeastern Ala., southern Ga., southern S.C., and coastal N.C.

68b Dorsum brownish with distinctly darker head, venter pale yellowish, greenish or white .... **Pine Woods Snake,**
Rhadinea flavilata (Cope)

Dorsum golden brown to reddish brown with darker brown head, upper lip white or yellow in Fla. populations, venter white to yellowish green; body length 25–40 cm; found along southeastern coastal plain from eastern La. to N.C., including northern two-thirds of peninsular Fla.

69a Dorsal color uniform green ....
.... **Smooth Green Snake,**
Opheodrys vernalis (Harlan)

Color is uniform green above, white or yellowish below; scales are smooth; body length 35–65 cm. Distribution (Fig. 464).

Figure 464

Figure 464   Distribution of *Opheodrys vernalis*

69b Dorsal color not green **70**

70a Venter with black crossbars, tail ends in sharp spine..
.... **Sharptail Snake,**
Contia tenuis Baird & Girard

Dorsum reddish brown or reddish gray, with faint reddish or yellowish lateral line, venter with black crossbars on gray or cream belly; body length 20–45 cm; found in mountains of northern half of Calif. northward into Oreg. with isolated populations in Wash.

70b Venter uniform light or any black consists of rings circling body **71**

71a Snout normal .... Ground Snake,
Sonora semiannulata Baird & Girard

Color pattern highly variable from uniform tan or reddish or with black or gray crossbands on anterior or entire length; body length 23–40 cm. Distribution (Fig. 465). Previously, some populations known as _Sonora episcopa_.

Figure 465

Figure 465   Distribution of _Sonora episcopa_

71b Snout flattened, shovel-like (Fig. 466); nasal valve present 72

72a Black mark on head forming a broad band across top of head ....
.... Sonoran Shovelnose Snake,
Chionactis palarostris Klauber

Figure 466

Figure 466   _Chionactis_ head shape

Color pattern of alternating red-orange and black bands, some encircling body, red band broader than in following species, ground color yellow;

body length 25–40 cm; found in desert of western Sonora, Mexico extending into extreme southern Ariz.

72b Black mark on head forming a crescent (extending between eyes) which is concave anteriorly ..... Western Shovelnose Snake,
Chionactis occipitalis (Hallowell)

The reddish orange bands are thinner than black bands, some bands may encircle body, ground color yellow; body length 25–43 cm; found in deserts from southern Nev. to southeastern Calif. and southern Ariz. and into Mexico adjacent to Colorado River.

73a (66) Rostral normal 76

73b Rostral enlarged with free lateral edges (Fig. 467) 74

74a Upper labials 8, posterior chin shields in contact or separated by a single small scale .... Mountain Patchnose Snake,
Salvadora grahamiae Baird & Girard

Figure 467

Figure 467   _Salvadora_

Middorsal light stripe (white, tan, gray, or yellowish) bordered by heavy dark brown to black stripes 2 + scales wide, Tex. subspecies with additional thin lateral dark line; body length 50–75 cm; found in southeastern Ariz., southern N. Mex., western central and south Tex. southward into Mexico.

74b Upper labials 9, 2–3 small scales between posterior chin shields **75**

75a Two upper labials reach eye, loreal single .... **Big Bend Patchnose Snake,** *Salvadora deserticola* **Schmidt**

Middorsal stripe of tan to brownish orange bordered by two thin lines of dark brown or black and a third laterally positioned dark line on scale row 3 or 4; body length 60–100 cm; extreme western Big Bend of Tex., southwestern N. Mex. and southeastern Ariz. southward into Mexico.

75b One or no upper labial reaches eye, loreal often divided... .... **Western Patchnose Snake,** *Salvadora hexalepis* **(Cope)**

Middorsal stripe of yellow or tan bordered by 2 or 3 thin dark brown to black lines; body length 50–115 cm. Distribution (Fig. 468).

Figure 468

Figure 468   Distribution of *S. hexalepis*

76a One anterior temporal, lower preocular normal **77**

76b Usually 2–3 anterior temporals, lower preocular wedged between upper labials **78**

77a Dorsal color green ... *Opheodrys vernalis,* **(see couplet 69a)**

77b Dorsal color not green, black ring on neck and/or black spots on belly .... **Ringneck Snake,** *Diadophis punctatus* **(Linnaeus)**

Dorsum dark brown, gray or bluish, with red or yellow ring around neck which may be incomplete or absent in some western populations, belly yellow to red; body length 25–50 cm. Distribution (Fig. 469).

Figure 469

Figure 469   Distribution of *D. punctatus*

78a Scale rows at posterior end of body 15 .... Racer, *Coluber constrictor* Linnaeus

Uniform black, blue, olive or brown above with black, yellow, or pale blue to cream belly depending on subspecies, young with pattern of blotches on light ground color; body length 75–160 cm. Distribution (Fig. 470). (Some authors consider western populations to be a distinct species *Coluber mormoni.*)

Figure 470

Figure 470   Distribution of *C. constrictor*

78b Scale rows at posterior end of body 11–13 79

79a Dorsal scale rows 15 at midbody .... Striped Whipsnake, *Masticophis taeniatus* (Hallowell)

Color black, reddish brown, greenish or bluish gray with longitudinal light lines on sides, belly variable from white, yellow to gray, venter of tail is orange or reddish; body length 100–200 cm. Distribution (Fig. 471).

Figure 471

Figure 471   Distribution of *M. taeniatus*

79b Dorsal scale rows 17 at midbody 80

80a No distinct longitudinal stripes ....
.... **Coachwhip,**
Masticophis flagellum **(Shaw)**

Dorsum uniform tan, brown, black or reddish or brown with blackish anterior, one subspecies bluish with light irregular placed small spots, belly color similar to back or lighter; body length 90–200 cm. Distribution (Fig. 472).

Figure 472

Figure 472 Distribution of *M. flagellum*

80b Distinct longitudinal stripes present ... **81**

81a A single lateral stripe which continues onto tail on each side of body ....
.... **Striped Racer,**
Masticophis lateralis **(Hallowell)**

Grayish black to dark brown with pale yellow or orangish stripe on each side, venter cream with underside of tail pink to coral; body length 75–150 cm; a snake of the chaparral along coastal Calif. and lower mountains surrounding the central valley and southward into Baja California.

81b 2–3 lateral stripes which do not continue onto tail .... **Sonora Whipsnake,**
Masticophis bilineatus **Jan**

Dorsum olive to bluish or brownish gray with 2–3 light stripes on each side which fade posteriorly, belly cream to pale yellow; body length 75–165

cm; a Mexican species whose northern distribution extends into central Ariz. and extreme southwestern N. Mex.

82a (63) Dorsal scale rows 13 ....
.... **Banded Sand Snake,**
Chilomeniscus cinctus **Cope**

Color pale yellow to orange with prominent black crossbands, belly light; body length 18–25 cm; found in southwestern Ariz. and southward into Mexico.

82b Dorsal scale rows more than 13 .... **83**

83a Dorsal scale rows 15 .... **87**

83b Dorsal scale rows 17 .... **84**

84a Head long with sharp nose (Fig. 473); tail long .... **Mexican Vine Snake,**
Oxybelis aeneus **(Wagler)**

Figure 473

Figure 473 *Oxybelis aeneus*

Very slender snake with gray to yellowish brown dorsum and gray to white belly; body length 100–150 cm; enters U.S. in Pajarito Mountains in extreme southern Ariz. and ranges southeasterly to Brazil.

84b Head shorter with rostral turned up at tip (Fig. 474); tail short and thick 85

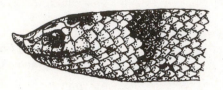

Figure 474

Figure 474 *Gyalopion* with turned-up rostral

85a Rostral separating small internasals and in contact with prefrontals 86

85b Rostral larger, separating prefrontals and in contact with frontal (no internasals) .... **Mexican Hooknose Snake,** *Ficimia streckeri* **Taylor**

Ground color of pale brown to gray with narrow brown or olive crossbands which may be reduced to rows of irregular spots; body length 18–35 cm; found in southern tip of Tex. and southward into Mexico.

86a Brown markings with black edges on head not forming large blotch (Fig. 475a) .... **Western Hooknose Snake,** *Gyalopion canum* **Cope**

Figure 475a

Figure 475a *Gyalopion canum*

Figure 475b

Figure 475b *Gyalopion quadrangularis*

Body gray to light brown with brown crossbands edged with black; body length 18–35cm; found in southwestern Tex., southern N. Mex. and southeastern Ariz.

86b Single large blotch on top of head and neck (Fig. 475b) .... Desert Hooknose Snake, *Gyalopion quadrangularis* (Gunther)

Dorsum white with reddish band at the middorsal line and prominent black dorsal saddles, belly greenish yellow; body length 15–30 cm; enters U.S. in extreme southern Ariz. in Santa Cruz County.

87a Snout flattened, shovel-like .. *Chionactis,* (see couplet 72)

87b Snout normal 88

88a Head is black 89

88b Head not distinctly black, may be slightly darker than body color .. Flathead Snake, *Tantilla gracilis* Baird & Girard

Dorsum light brown (reddish to grayish) head slightly darker, belly salmon pink; body length 18–20 cm. Distribution (Fig. 476).

Figure 476

Figure 476   Distribution of *Tantilla gracilis*

89a Light colored (usually white) crossband or collar on back of head or neck 90

89b No such crossband or collar 96

90a White color touching back of head scales or including posterior edge of head scales (Fig. 477a) 93

90b Collar on neck not touching head scales (Fig. 477b) 91

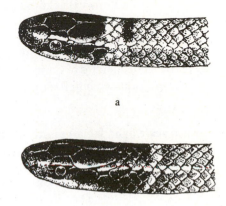

a

b

Figure 477

Figure 477   Collar variation in *Tantilla*

91a Black of head extends to top of labials and below corner of mouth 92

91b Cream colored spot on side of head .... Yaqui Blackhead Snake, *Tantilla yaquia* Smith

Light brown to grayish with black head and white ring on neck; body length 18–35 cm; range extends into southeastern Ariz. from Mexico.

92a Count of ventral and caudal scales usually greater than 245 .... **Desert Blackhead Snake,** *Tantilla transmontana* **Klauber**

Color brown to olive gray with black head and narrow white collar, orange to reddish stripe down venter; body length 18–35 cm; distribution restricted to southern Calif. from Riverside to San Diego Counties.

92b Count of ventral and caudal scales usually less than 245................................ .... **California Blackhead Snake,** *Tantilla eiseni* **Stejneger**

Color like *T. transmontana;* body length 18–35 cm; found in coastal mountains of California from just south of San Francisco into Baja California.

93a Rostral scale light, labials mostly black or with light spot in middle of series, collar may be broken with black at middorsal line .... **Big Bend Blackhead Snake,** *Tantilla rubra* **Cope**

Dorsum light brown to grayish brown, black head with or without white collar; body length 22–45 cm; found in scattered locations in Big Bend and Devils River area of southwestern Tex.

93b Rostral scale black or light color extending to cover most of internasals **94**

94a Light spot on rostral and internasals or if black, then all of upper labials are black .... **Florida Crowned Snake,** *Tantilla relicta* **Telford**

Color light brown, black head with or without light collar; body length 18–23 cm; distributed across central Fla. peninsula, but not in extreme southern Fla. (see *Tantilla oolitica,* couplet 97b).

94b Rostral black and at least part of upper labials not black **95**

95a Light collar followed by black band 3–5 scales wide................................ .... **Southeastern Crowned Snake,** *Tantilla coronata* **Baird & Girard**

Dorsum light brown to reddish brown, black head with light collar followed by broad black band; body length 20–30 cm. Distribution (Fig. 478).

Figure 478

Figure 478   Distribution of *T. coronata*

95b Light collar followed by faint black band 1/2 to 1 1/2 scales wide .... **Chihuahuan Blackhead Snake,** *Tantilla wilcoxi* **Stejneger**

Dorsum light brown, black cap bordered by broad white collar that crosses tip of parietals; body length 18–35 cm; its Mexican distribution extends into extreme southeastern Ariz. in Huachuca and Patagonia Mountains.

96a (89) Black of head extending minimally 4–6 scale rows beyond head **97**

96b Black of head extending 1–3 scale rows beyond head **98**

97a All labials black .... *Tantilla rubra,*
(see couplet 93)

97b Some of upper labials not black................
Tantilla relicta, (see couplet 94) ....
.... **Rim Rock Crowned Snake,**
Tantilla oolitica **Telford**

Color brown, black head extending several scales beyond head; body length 18–25 cm; restricted to extreme southern tip of Fla. (Dade County and Key Largo).

98a Black head cap convex posteriorly ....
.... **Plains Blackhead Snake,**
Tantilla nigriceps **Kennicott**

Dorsum yellowish brown to brownish gray, black head cap convex behind; body length 18–30 cm. Distribution (Fig. 479).

Figure 479

Figure 479   Distribution of *T. nigriceps*

98b Black head cap not convex posteriorly
.... **99**

99a Fewer ventral scales (less than 152 in males and less than 160 in females) ....
.... **Mexican Blackhead Snake,**
Tantilla atriceps **(Günther)**

Brown to grayish brown above with bright orange-red band on belly, black head extends just beyond parietals; body length 13–22 cm; found from southwestern Tex. to central Ariz. and southward into Mexico.

99b More ventral scales (greater than 152 in males and greater than 161 in females) ...
.... **Utah Blackhead Snake,**
Tantilla utahensis **Blanchard**

Brownish dorsum, black cap as above, light collar rare; body length 18–38 cm; spotty distribution from Inyo Co., Calif., across Nev. into southern Utah.

100a (62) No preocular (Fig. 480a) **101**

100b One or more preoculars present (Fig. 480b)
.... **102**

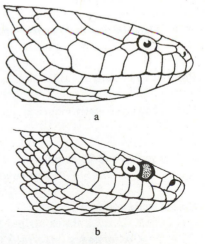

a

b

Figure 480

Figure 480   Preocular absent (a) or present (b)

101a (11, 13) Dorsal pattern with longitudinal stripes .... Rainbow Snake, *Farancia erythrogramma* (Latreille)

Pattern of longitudinal red stripes on black, belly with small black spots bordering central red band; body length 90–140 cm; found from southern Md. to central Fla. and eastern La.

101b No longitudinal stripes .... Mud Snake, *Farancia abacura* (Holbrook)

Dorsum black with red or pink blotches along ventrolateral aspect, belly red with large black spots alternating along sides; body length 100–150 cm. Distribution (Fig. 481).

Figure 481

Figure 481   Distribution of *Farancia abacura*

102a Pupil of eye round 103

102b Pupil of eye vertically elliptical 104

103a Dorsal scale rows 25 or more ..... *Elaphe*, (see couplet 27)

103b Dorsal scale rows 19 .... Black-striped Snake, *Coniophanes imperialis* (Baird)

Dorsal pattern of broad black or dark brown stripes, alternating with tan or brown stripes,

venter bright red or orange; body length 30–50 cm; extreme southern tip of Tex. southward into Middle America.

104a 1 loreal, 7 or 8 upper labials 105

104b 2 or more loreals, usually 9 or more labials .... *Trimorphodon biscutatus*, (see couplet 51b)

105a Dorsal pattern of many small to medium dark brownish spots on lighter ground color .... Night Snake, *Hypsiglena torquata* Gunther

Elongated dark blotch on each side of neck, venter white or yellowish; body length 35–45 cm. Distribution (Fig. 482).

Figure 482

Figure 482   Distribution of *Hypsiglena torquata*

105b Dorsal pattern, series of large dark saddles on light ground color (usually yellow) .... Cat-eyed Snake, *Leptodeira septentrionalis* (Kennicott)

Dorsal color yellow to tan with broad dark brown to black saddles; body length 45–90 cm; found in extreme southern Tex. southward to Veracruz, Mexico.

CROCODILES AND ALLIGATORS
Crocodylia

Crocodilians are large lizard-like reptiles with strong muscular tails, formidable jaws, and a thick leathery skin reinforced with bony plates in the lower dermal layer. Because of their conspicuous size and potential danger to man, alligators and crocodiles are familiar subjects to even the most biologically naive. However, only two species (Nile Crocodile and Salt Water Crocodile of Africa and India) are considered regularly dangerous to man although large individuals of other species are dangerous if cornered or threatened. Also because of their conspicuous size, distinct habitat requirements, and commercially valuable skin, man and his activities have greatly reduced natural populations and many species are threatened with extinction.

Modern crocodilians (8 genera and 4 families) represent only 1 group of 5 major lineages present in Mesozoic times and are considered to be descended from the archosaurian reptiles which gave rise to modern birds. Crocodilians have a diapsid skull (a dorsal and temporal opening completely bounded by arches on both sides of the posterior skull), a clear membrane (nictitating) which covers the eye, special abdominal ribs (gastralia) derived from the dermal armor of the skin, skin reinforced by bony plates, and a longitudinal cloacal opening and no urinary bladder. Males have a single penis like turtles rather than paired hemipenes as in lizards and snakes. Crocodiles represent the reptilian group most similar to homeotherms because of the four-chambered heart, thecodont teeth, (set in sockets), true cerebral cortex (well developed outer layer of the anterior brain), and a secondary palate which serves to separate the air passage with the food canal in the mouth. The latter characteristic allows crocodiles and alligators to sit submerged in water with only the eyes and nostrils protruding as they wait for prey. The mouth is not water tight but water is prevented from entering the lungs by a valve at the back of the mouth.

Ecology of natural populations has not been extensively studied but in general crocodilians require several years to mature (8–10 yrs.) and may live to be 50–60 years old growing about a foot per year until maturity and then at a much slower rate. Presumably only a single clutch of eggs numbering 20–100 are laid in a nest of sand or vegetation about 2 feet deep. The eggs are 2–3 inches long and have hard calcified shells. The female guards the nest in many species and has been observed to dig it up upon hearing the croaking sounds of newly hatched young. All crocodilians are capable of producing sound which varies from a moaning grunt to a gutteral bark. The American Alligator is particularly noted for its loud and penetrating voice.

The jaws are lined with large teeth which are used to seize food rather than grind it. Food habits vary with size of the individual and range from insects and frogs in hatchlings to fish, clams, birds, and mammals in larger individuals. Food is swallowed whole with larger items being torn into fragments with a violent twisting and rolling motion of the head and body. The stomach is muscular and serves in the same way as the gizzard of birds. Rocks and other solid objects are swallowed to aid this grinding.

Crocodilians occur around the world in tropical and subtropical aquatic habitats in rivers, lakes, marshes, and swamps. Some species occur principally in fresh water and others in salt or brackish water but all are closely associated with permanent water. Only two species (*Alligator mississippiensis* and *Crocodylus acutus*) are native to the U.S. but a third species (*Caiman crocodylus*) is occasionally encountered because of the pet trade and thus is included here.

1a Transverse bony ridge in front of eyes (Fig. 483) .... Spectacled Caiman, *Caiman crocodylus* (Linnaeus)

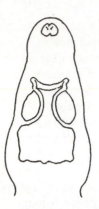

Figure 483

Figure 483   *Caiman crocodylus*

Native from southern Mexico to Argentina; commonly sold in pet stores and liberated specimens occasionally encountered in the wild in the southeastern U.S.; total length 20 cm at hatching and largest adults about 245 cm. Color is gray (variously tinted green, yellow or brown) with dark brown crossbands.

1b No transverse bony ridge in front of eyes 2

2a Snout long and tapering about 1/4 as wide at tip as at base of jaw, (Fig. 484) .... American Crocodile, *Crocodylus acutus* Cuvier

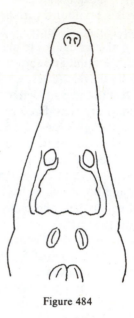

Figure 484

Figure 484   *Crocodylus acutus*

Natural populations rare in U.S. in extreme southern tip of Fla., Florida Keys, Everglades National Park, disjunct range through Carribean Islands, southern Mexico to Colombia and Ecuador; adults 225 cm to 455 cm total length in U.S., up to 700 cm. in South America; color is gray (tan or green tinted) with darker crossbands or blotches; young gray with distinct narrow black crossbands of spots.

2b Snout blunt with tip about 1/2 as wide as skull at base of jaw (Fig. 485) upper teeth overlap mandibular teeth and 4th mandibular fits into a socket of upper jaw, not protruding outside of jaw .... **American Alligator,** *Alligator mississippiensis* **(Daudin)**

introduced in lower Rio Grande valley of Tex.; adults 180–580 cm total length; color black with lighter markings in younger individuals; hatchlings 23 cm total length, black with distinct yellow crossbands.

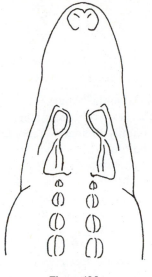

Figure 485

Figure 485   *Alligator mississippiensis*

Distributed in southeastern U.S. along Gulf and Atlantic coastal areas from N.C. to central Tex.,

Classification Outline

Amphibians and reptiles belong to the Phylum Chordata and Subphylum Vertebrata. The classification followed in this book is outlined below with genera of U.S. species given for each family.

Class Amphibia
  Superorder Lissamphibia
    Order Caudata (Urodela)
      Suborder Cryptobranchoidea
        Family Cryptobranchidae—
          Cryptobranchus
      Suborder Salamandroidea
        Family Ambystomatidae—
          Ambystoma, Rhyacotriton
        Family Dicamptodontidae—*Dicamptodon*
        Family Amphiumidae—*Amphiuma*
        Family Proteidae—*Necturus*
        Family Salamandridae—*Notophthalmus, Taricha*
        Family Plethodontidae—*Aneides, Batrachoseps, Desmognathus, Ensatina, Eurycea, Gyrinophilus, Haideotriton, Hemidactylium, Hydromantes, Leurognathus, Phaeognathus, Plethodon, Pseudotriton, Stereochilus, Typhlomolge*
    Suborder Meantes
      Family Sirenidae—
        Pseudobranchus, Siren

Order Salientia (Anura)
  Suborder Archeobatrachia
    Family Pipidae—*Xenopus*
    Family Rhinophrynidae—
      Rhinophrynus
    Family Ascaphidae—*Ascaphus*
    Family Pelobatidae—*Scaphiopus*
  Suborder Neobatrachia
    Family Leptodactylidae—
      Eleutherodactylus, Hylactophryne, Leptodactylus, Syrrhophus
    Family Bufonidae—*Bufo*
    Family Hylidae—*Acris, Hyla, Limnaoedus, Osteopilus, Pseudacris, Pternohyla, Smilisca*
    Family Microhylidae—
      Gastrophryne, Hypopachus
    Family Ranidae—*Rana*
Class Reptilia
  Subclass Anapsida
    Order Chelonia (Testudines)
      Suborder Cryptodira
        Superfamily Testudinoidea
          Family Chelydridae—*Chelydra, Macroclemys*
          Family Emydidae—*Chrysemys, Clemmys, Deirochelys, Emydoidea, Graptemys, Malaclemys, Terrapene*

Family Kinosternidae—
    Kinosternon, Sternotherus
Family Testudinidae—*Gopherus*
Superfamily Trionychoidea
    Family Trionychidae—*Trionyx*
Superfamily Chelonioidea
    Family Cheloniidae—*Caretta,*
      Chelonia, Eretmochelys,
      Lepidochelys
    Family Dermochelyidae—
      Dermochelys

Subclass Lepidosauria
  Order Squamata
    Suborder Lacertilia (Sauria)
      Infraorder Gekkota
        Family Gekkonidae—*Anarbylus,*
          Coleonyx, Gonatodes,
          Hemidactylus, Phyllodactylus,
          Sphaerodactylus
      Infraorder Iguania
        Family Iguanidae—*Anolis,*
          Callisaurus, Cophosaurus,
          Crotaphytus, Dipsosaurus,
          Gambelia, Holbrookia,
          Leiocephalus, Petrosaurus,
          Phrynosoma, Sauromalus,
          Sceloporus, Uma, Urosaurus,
          Uta
      Infraorder Anguimorpha
        Family Anguidae—*Elgaria,*
          Gerrhonotus, Ophisaurus
        Family Helodermatidae—
          Heloderma
        Family Anniellidae—*Anniella*
      Infraorder Scincomorpha
        Family Scincidae—*Eumeces,*
          Neoseps, Scincella
        Family Teiidae—*Cnemidophorus*
        Family Xantusiidae—*Klauberina,*
          Xantusia

Suborder Amphisbaenia
    Family Rhineuridae—*Rhineura*
Suborder Serpentes (Ophidia)
  Infraorder Scolecophidia
    Family Leptotyphlopidae—
      Leptotyphlops
  Infraorder Henophidia
    Family Boidae—*Charina,*
      Lichanura
  Infraorder Caenophidia
    Family Micruridae—*Micruroides,*
      Micrurus
    Family Viperidae—*Agkistrodon,*
      Crotalus, Sistrurus
    Family Colubridae—*Arizona,*
      Carphophis, Cemophora,
      Chilomeniscus, Chionactis,
      Clonophis, Coluber, Contia,
      Diadophis, Drymarchon,
      Drymobius, Ela-
      phe, Farancia, Ficimia,
      Gyalopion, Heterodon,
      Hypsiglena, Lampropeltis,
      Leptodeira, Masticophis,
      Nerodia, Opheodrys, Oxybelis,
      Phyllorhynchus, Pituophis,
      Regina, Rhadinaea,
      Rhinocheilus, Salvadora,
      Seminatrix, Sonora, Stilosoma,
      Storeria, Tantilla,
      Thamnophis, Trimorphodon,
      Tropidoclonion, Virginia
Subclass Archosauria
  Order Crocodylia
    Family Crocodylidae—*Alligator,*
      Caiman, Crocodylus

Index
and Pictured Glossary

(a)

(b)

Fig. 486 Inguinal (a) and axillary (b)
amplexus

viz., anapsid (no openings), euryapsid, and synapsid (each one opening). Second, used in the sense of diapsid 'reptiles', that is all groups having the anatomical condition described above (snakes, lizards, amphisbaenians, tuataras, crocodiles, and birds) although sometimes in modified conditions. Fig. 487, 213

Fig. 487 Diapsid Skull

PARAPHYLETIC GROUP: Such groups are taxonomist's illusions. Like monophyletic groups, all members of the paraphyletic group are hypothesized to be descendants of a single common ancestral species but proponents of paraphyletic groups do not require that all descendants of that common ancestor be members of the group. Any two species can form a paraphyletic group—such groups, if characterizable, are characterized by PLESIOMORPHIC features, 1

parasphenoid teeth, 17, fig. 14a, 31

parietal, 122

parietal eye, 122, 124, 166

PAROTOID GLANDS: Enlarged glands found posterior to the eyes, most frequently in true toads. These glands secrete offensive and/or toxic substances when the toads are severely molested, 16, fig. 12, 26, fig. 41, 53, fig. 90

Patchnose Snake, Big Bend, 205
  Mountain, 204, fig. 467
  Western, 205, fig. 468

PECTORAL GIRDLE: "Shoulder bones" attaching anterior limbs to body, 54

pectoral shield, 54, 95, fig. 191b

PELAGIC: Living in open waters such as the sea turtles or amphibian larvae, 58, 103

Pelobatidae, 55, 60, 92, 216

Pelobatid larva, 92, fig. 183

Pelomedusidae, 7

PELVIC GIRDLE: "Hip bones" attaching posterior limbs to body, 168

PERENNIBRANCHS: Organisms with permanent gills. Neotenic salamanders include several taxa which retain the gills and complete gill arches as adults (e.g., proteids, sirenids, and some plethodontids and some ambystomatids), 16

perstriatus, Notophthalmus, 32, fig. 55d

Petrosaurus, 142, 217
  mearnsi, 142, fig. 305

Phaeognathus, 43, 216
  hubrichti, 43

PHALANX: A digit of either the hand or foot, 54

pholeter, Amphiuma, 28

Phrynosoma, 129, 130, 131, 132, 217
  cornutum, 130, fig. 279, 131, fig. 282, 283
  douglassi, 131, fig. 281
  mcalli, 131
  modestum, 130, fig. 280
  platyrhinos, 132, fig. 284
  solare, 129, fig. 278

Phyllodactylus, 128, fig. 275, 217
  xanti, 128, fig. 276

Phyllorhynchus, 180, 192, fig. 435, 199, 217
  browni, 199, fig. 454
  decurtatus, 199, fig. 453

picta, Chrysemys, 112, fig. 241, fig. 242

Pig Frog, 83

Pigmy Rattlesnake, 173, fig. 379

Pigmy Salamander, 44

Pine Barrens Treefrog, 79

Pine Snake, 192, fig. 434

Pine Woods Snake, 203

Pine Woods Treefrog, 79

pipid larva, 92

Pipidae, 4, 55, 57, 58, 92, 216

piscivorus, Agkistrodon, 172

Pit-fall traps, 10, fig. 9b

Pit Vipers, 171

Pituophis, 179, 217
  catenifer, 191, fig. 432, fig. 433
  melanoleucas, 192, fig. 434

Plainbelly Water Snake, 189, fig. 424a, fig. 425

Plains Blackhead Snake, 211

Plains Garter Snake, 194, fig. 441

Plains Spadefoot, 62

planirostris, Eleutherodactylus, 64, fig. 121a

plastron, 95, fig. 191b

Plateau Spotted Whiptail, 161, fig. 354b

Plateau Striped Whiptail, 164, fig. 360

platineum, Ambystoma, 27

platycephalus, Hydromantes, 42

platyrhinos, Heterodon, 181, fig. 402

Platysternidae, 7

Platysternon macrocephalus, 7

PLESIOMORPHIC: The primitive condition within a group. Plesiomorphic similarities depend on having identified a monophyletic group and one or more "highly" derived members of it as a second coordinate group. The second group is characterized by unique derived features but the other members of the first monophyletic group do not constitute a logically defined group even though they can be collectively characterized by the primitive condition, 29

Plethodon, 46, 48, 49, 50, 51, 53, 216
  caddoensis, 51
  cinereus, 52, 53, fig. 88, fig. 89
  dorsalis, 53
  dunni, 48, fig. 84a, 49, fig. 85
  elongatus, 50
  glutinosus, 50, fig. 87
  gordoni, 48
  hoffmani, 52
  jordani, 51
  larselli, 49
  longicrus, 51
  neomexicanus, 50
  nettingi, 52
  ouachitae, 51
  punctatus, 52
  richmondi, 51
  serratus, 53
  stormi, 50
  vandykei, 48
  vehiculum, 49, fig. 84b
  wehrlei, 52
  welleri, 50
  yonahlosee, 51

Plethodontidae, 4, 17, fig. 14a, 18, 19, 31, 34, 35, 91, 216

Pleurodira, 7, 95

PLEURODONT: Condition where teeth are attached to the inside surface of the jaw bone, see TOOTH ATTACHMENT, fig. 488

POIKILOTHERMIC: "Cold blooded", animals which are unable to maintain a constant body temperature by internal metabolic processes, 166

poinsetti, Sceloporus, 138, fig. 298

polyphemus, Gopherus, 102, fig. 209

Pond Turtle, Western, 116

Pond Turtle Family, 109

porphyriticus, Gyrinophilus, 40, fig. 70

postantebrachial scales, 163

POSTERIOR: End of body away from the head and toward the tail, 139, 148

posterior antebrachials, 123, 267

postmental, 122, fig. 266, 152, fig. 329

postnasal, 150, fig. 322a, 173, fig. 379

postoculars, 166

postrostral, 122, fig. 266, 155, fig. 337

Prairie Skink, 153, fig. 331, fig. 332

preanal scales, 163, fig. 359

prefrontal, 122, fig. 266, 166

PREMAXILLA: The anterior most bone of the maxillary arch; normally paired and bearing teeth, 5

preocular, 166, 211, fig. 480

pretiosa, Rana, 86, fig. 167

pricei, Crotalus, 174, fig. 383

propinqua, Holbrookia, 133

Proteidae, 16, 19, 29, 90, 216

Proteus, 29

proximus, Thamnophis, 195, fig. 443

Pseudacris, 76, 216
  brachyphona, 77
  brimleyi, 76
  clarki, 77
  nigrita, 77
  ornata, 77
  streckeri, 77, fig. 148
  triseriata, 76, fig. 146, fig. 147

Pseudidae, 4, 73

Pseudobranchus, 33, 216
  striatus, 33

pseudogeographica, Graptemys, 117, fig. 260

Pseudotriton, 41, fig. 71, 91, 216
  montanus, 41, fig. 72
  ruber, 41, fig. 71

Pternohyla, 73, 216
  fodiens, 73, fig. 140

pulchra, Anniella, 122, 156, fig. 341

pulchra, Graptemys, 117, 118, fig. 261a, fig. 262

punctatus, Bufo, 68, fig. 127, fig. 128

punctatus, Diadophis, 205, fig. 469

punctatus, Necturus, 30

punctatus, Sphenodon, 6, 120

pygaea, Seminatrix, 203

Pygopodidae, 7

pyromelana, Lampropeltis, 201

Python Family, 169

Q

quadramaculatus, Desmognathus, 45, fig. 80

quadrangularis, Gyalopion, 208, fig. 475b, 209

quadridigitata, Eurycea, 36, fig. 64

Queen Snake, 186, fig. 416a, fig. 417

quercicus, Bufo, 70

R

Racer, 11, 206
  Speckled, 185
  Striped, 207

Racerunner, Six-lined, 163, fig. 357, fig. 358

radix, Thamnophis, 194, fig. 441

Rainbow Snake, 212

Rana, 53, fig. 91, 82, 83, 84, 85, 86, 87, 216
  areolata, 84, fig. 163a, fig. 164
  aurora, 86
  berlandieri, 89
  blairi, 89, fig. 174, fig. 175
  boyli, 83
  cascadae, 85, fig. 163b
  catesbeiana, 83, fig. 159, fig. 160
  chiricahuensis, 89
  clamitans, 84, fig. 161, fig. 162
  fisheri, 89
  grylio, 83
  heckscheri, 82
  muscosa, 83
  onca, 89
  palustris, 87, fig. 169, fig. 170
  pipiens, 82, 88, fig. 171, fig. 172
  pretiosa, 86, fig. 167, 89
  septentrionalis, 82
  sylvatica, 85, fig. 165, fig. 166
  tarahumarae, 83
  utricularia, 88, fig. 173
  virgatipes, 82

Ranidae, 56, 82, 93, 216

rathbuni, Typhlomolge, 38, fig. 67

Rat Snake, 11, 191, fig. 430
  Green, 190
  Trans-Pecos, 190, fig. 427

Rattlesnakes, 171
  Blacktail, 178, fig. 395
  Eastern Diamondback, 178, fig. 393, fig. 394
  Mohave, 177, fig. 389
  Pigmy, 173, fig. 379a
  Red Diamond, 177, fig. 390a
  Ridgenose, 174, fig. 382
  Rock, 175, fig. 386
  Speckled, 176, fig. 388
  Tiger, 176, fig. 387
  Timber, 179, fig. 396
  Twin-spotted, 174, fig. 383
  Western, 175, fig. 384, fig. 385
  Western Diamondback, 177, fig. 390b, fig. 391, 178, fig. 392

Ravine Salamander, 51

Razorback Musk Turtle, 107

Rear-fanged snakes, 179

Redback Salamander, 52

Redbelly Snake, 182, fig. 406, 183, fig. 407

Redbelly Turtle, 114
  Alabama, 114
  Florida, 114

Red Diamond Rattlesnake, 177, fig. 390a

Red Eft, 31

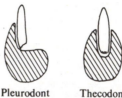